PRACTICAL ENGLISH in OFFSHORE PLANT ENGINEERING

DONG H. KIM

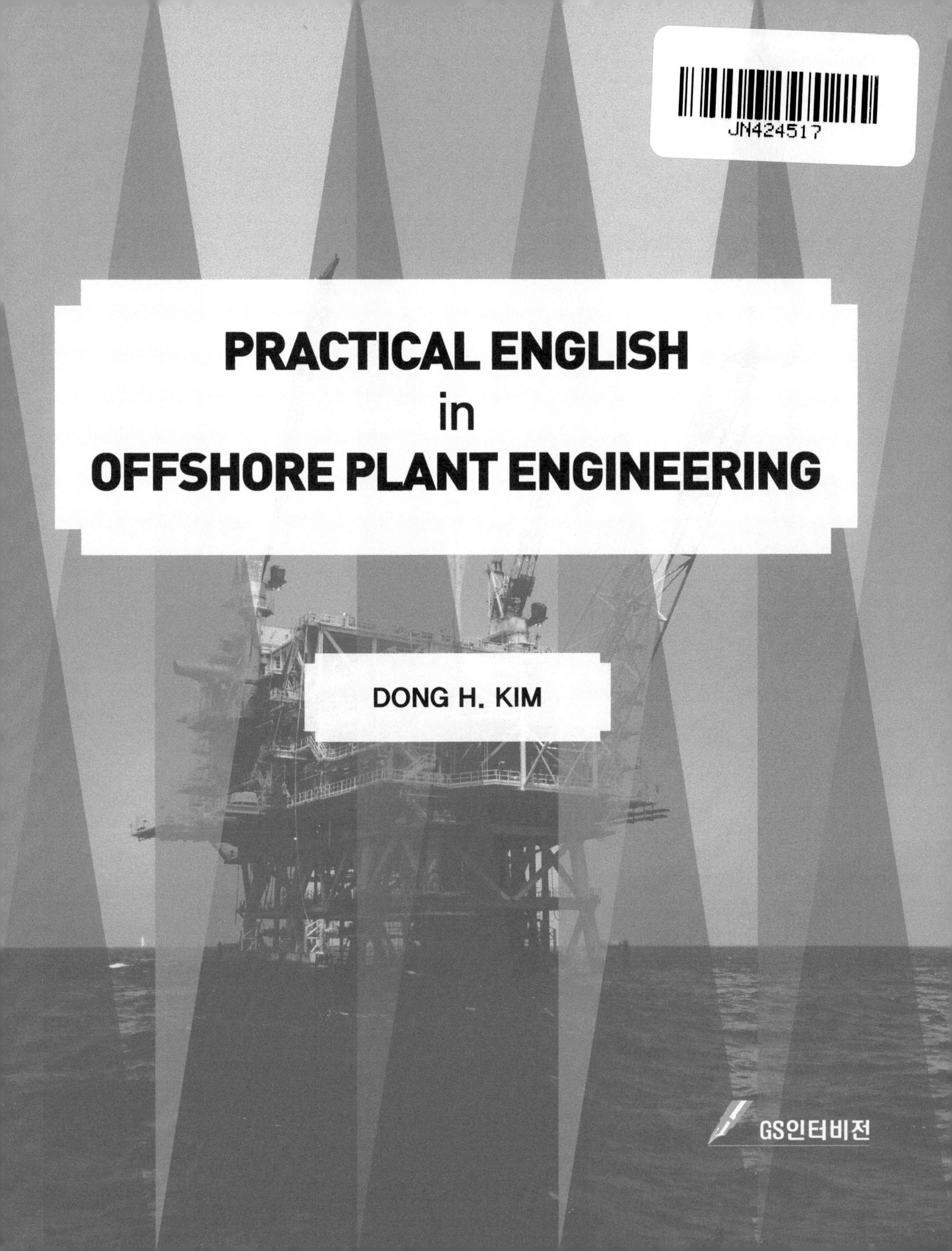

GS인터비전

머리글

우리나라가 산업선진국의 대열에 합류하게 된 것은 불과 수 년 밖에 되지 않지만 조선산업의 경우 70년대 중반 들어 현대조선이 설립된 이래 이미 80년 대부터는 삼성중공업 및 대우중공업 등의 대형 조선소의 덕분에 조선강국으로 자리잡게 되었습니다. 1990년대에 들어서는 일본과 세계시장을 양분하더니 급기야는 2000년대 들어 일본을 추월하고 세계제일의 조선국가로 발돋움하였습니다. 그러나 이러한 조선1위는 중국의 등장으로 위협받고 있으며 이미 선박의 건조 물량에서는 추월을 당하고 있는 실정입니다. 결국 우리나라의 조선산업은 구조조정 등을 통해 생존을 모색하고 있으나 궁극적으로는 개술개발을 통해 경쟁력을 키우는 것만이 해법이라고 보여집니다.

국내 조선산업들이 앞으로 해양플랜트 건조기술을 개발하는데 중점을 두고자 하겠다는 것은 이러한 관점에서 볼 때 시의 적절한 것이 아닌가 생각됩니다. 근해 지역에 부존하는 해양자원들이 점점 고갈되어감에 따라 인류는 점점 더 심해저의 개발에 의존할 수 밖에 없으며 심해저개발에 반드시 필요한 해양플랜트 건조기술이야말로 우리가 중점적으로 개발해야 할 분야일 것입니다.

이 책은 해양플랜트 건조기술과 관련된 지식을 영어로 공부하도록 만들어졌습니다. 해양플랜트 건조기술이 유럽과 미국 등 선진국에서 유래했고 고객들 또한 그들이기 때문에 관련지식을 영어로 배우는 것은 필수적일 것입니다. 이 책에서 해양플랜트 건조기술과 관련된 모든 분야를 망라할 수는 없었습니다마는 중요한 부분에 대한 예문들을 엄선하여 수록하였기 때문에 해양플랜트 산업에서 사용되는 수많은 용어들을 접할 수 있는 좋은 기회가 될 것이라고 자부합니다.

이 책을 공부하면서 얻게 되는 부수적인 효과는 전공관련 지식을 함께 습득할 수 있다는 점과 대학생들의 취업에서 중요한 역할을 하는 영어구사능력을 향상시킬 수 있다는 점입니다. 해양플랜트 건조기술이 기계공학, 화학공학, 조선공학 등의 전공지식을 많이 응용하기 때문에 관련 전공지식의 함양에도 도움이 되며, 동

시에 영문독해 능력을 향상시켜 TOEIC 등과 같은 인증시험에서 향상된 점수를 취득할 수 있을 것입니다.

매 unit 마다 영어로 된 해양플랜트 관련 문장들을 공부한 후에는 Home Work Problem을 통해

책에 있는 내용을 익히고 단어 및 전문용어들을 습득할 수 있도록 하였습니다. 처음에는 다소 생소한 분야라서 공부하기 힘들 수 있으나 곧 적응이 되어 원만한 학습성과를 올릴 수 있을 것이라 생각합니다. 아무쪼록 이 책을 잘 활용하여 좋은 성과를 거둘 수 있기를 기원합니다.

2014년 8월 23일 아치섬에서 저자

Contents

UNIT 4

UNIT 5

UNIT 6

UNIT 7

UNIT 13

UNIT 1

INTRODUCTION TO OFFSHORE PLANT ENGINEERING

1.1 Introduction

The offshore industry requires continued development of new technologies in order to produce oil in regions, which are inaccessible to exploit with the existing technologies.

Sometimes, the cost of production with the existing know-how makes it unattractive. With the depletion of onshore and offshore shallow water reserves, the exploration and production of oil in deep water has become a challenge to the offshore industry. Offshore exploration and production of minerals is advancing into deeper waters at a fast pace. Many deepwater structures have already been installed worldwide. New oil/gas fields are being discovered in ultra-deep water. Many of these fields are small and their economic development is a challenge today to the offshore engineers. This has initiated the development of new structures and concepts. Many of these structures are unique in many respects and their efficient and economic design and installation are a challenge to the offshore community. This will be discussed in more detail later. In order to meet the need for offshore exploration and production of oil/gas, a new generation of bottom-supported and floating structures is being developed.

The purpose of this chapter is to introduce the historical development of offshore structures in the exploration of petroleum reservoirs below the seafloor.

The chapter covers both the earlier offshore structures that have been installed in shallow and intermediate water depths and the various concepts suitable for deep-water development as well as those proposed as ultra-deep water structures. A short description of these structures is given and their applications are discussed.

1.2 Definition of Offshore Structures

An offshore structure has no fixed access to dry land and may be required to stay in position in all weather conditions. Offshore structures may be fixed to the seabed or may be floating. Floating structures may be moored to the seabed, dynamically positioned by thrusters or may be allowed to drift freely. The engineering of structures that are mainly used for the transportation of goods and people, or for construction, such as marine and commercial ships, multi-service vessels (MSVs) and heavy-lift crane vessels (HLCVs) used to support field development operations as well as barges and tugs are not discussed in detail in this book. While the majority of offshore structures support the exploration and production of oil and gas, other major structures, e.g. for harnessing power from the sea offshore bases, offshore airports are also coming into existence. The design of these structures uses the same principles as covered in this book, however they are not explicitly included herein.

We focus primarily on the structures used for the production, storage and offloading of hydrocarbons and to a lesser extent on those used for exploration.

1.3 Historical Development

The offshore exploration of oil and gas dates back to the nineteenth century. The first offshore oil wells were drilled from extended piers into the waters of Pacific Ocean offshore Summerlands, California in the 1890s (and offshore Baku, Azerbaijan in the Caspian Sea). However, the birth of the offshore industry is commonly considered as in 1947 when Kerr-McGee completed the first successful offshore well in the Gulf of Mexico in 15 ft (4.6 m) of water off Louisiana.

The drilling derrick and draw works were supported on a 38 ft by 71 ft (11.6 m by 21.6 m) wooden decked platform built on 16 24-in. (61-cm) pilings driven to a depth of 104 ft (31.7 m). Since the installation of this first platform in the Gulf of Mexico over 50 years ago, the offshore industry has seen many innovative structures, fixed and floating, placed in progressively deeper waters and in more challenging and hostile environments. By 1975, the water depth extended to 475 ft (144 m). Within the next three years the water depth dramatically leapt twofold with the installation of COGNAC platform that was made up of three separate structures, one set on top of another, in 1025 ft (312 m). COGNAC held the world record for water depth for a fixed structure from 1978 until 1991. Five fixed structures were built in water depth greater than **1000 ft** (328 m) in the 1990s. The deepest one of these is the Shell Bullwinkle platform in 1353 ft (412 m) installed in 1991. The progression of fixed structures into deeper waters up to 1988 is shown in fig.1.1

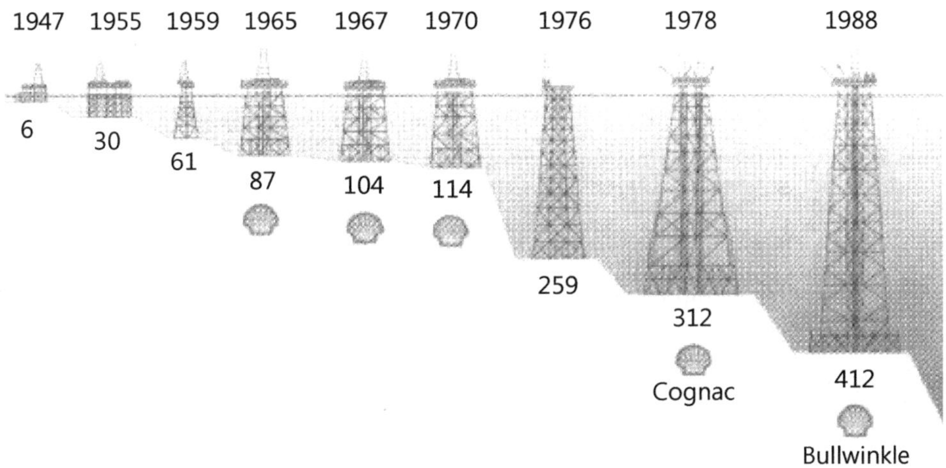

▲ **Figure 1.1** Progression of fixed platforms in the GOM – depths in meters (Courtesy Shell)

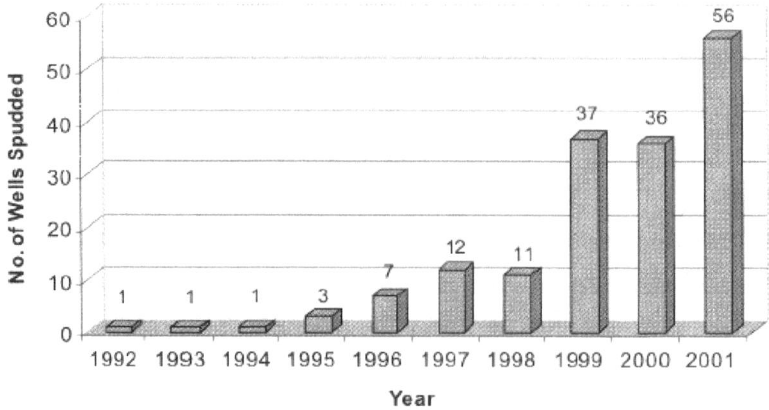

▲ **Figure 1.2** Ultra-deep water (>5000ft or 1524m) wells drilled in the Gulf of Mexico

Since 1947, more than 10,000 offshore platforms of various types and sizes have been constructed and installed worldwide. As of 1995, 30% of the world's production of crude came from offshore. Recently, new discoveries have been made in increasingly deeper waters. In 2003, 3% of the world's oil and gas supply came from deepwater (>1000 ft or 305 m) offshore production. This is projected to grow to 10% in the next fifteen years. The bulk of the new oil will come from deep and ultra-deepwater production from three offshore areas, known as the "Golden Triangle", the Gulf of Mexico, West Africa and Brazil. Figure 1.2 illustrates the recent growth in ultra-deepwater drilling in the Gulf of Mexico. Drilling activity is indicative of future production.

The importance of deepwater production to the US is illustrated in fig.1.3. US oil production is on the decline, dropping from about 7.5 MM BPD in 1989 to 5.9 MM BPD in 2001. The current US oil consumption is about 20 MM BPD. Experts do not believe there are significant new resources onshore in the US. Deepwater production has grown from 9.5% of US production in 1989 to 26.4% in 2001 (from 750,000 to 1,500,000 BPD). The drilling activity shown in fig.1.2 suggests that this percentage will continue to grow.

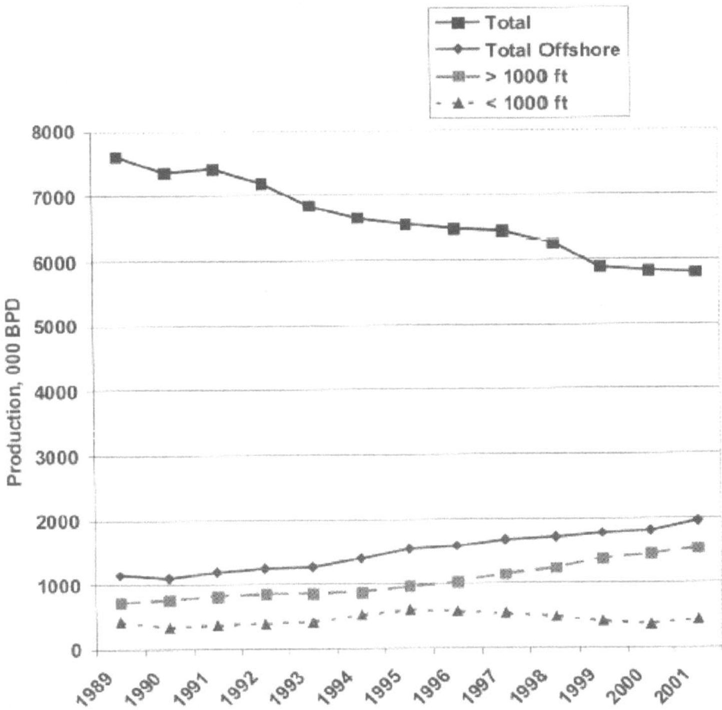

▲ **Figure 1.3** US crude oil production trends : importance of deep water [Source : Westwood(www.dw-I.com and OG) Database(www.ogi.com)]

Fixed structures became increasingly expensive and difficult to install as the water depths increased. An innovative and cheaper alternative to the fixed structure, namely, the Lena guyed tower was introduced in 1983. The platform was built in such a way that the upper truss structure could deflect with the wave and wind forces. Piles extending above the sea floor could bend, and horizontal mooring lines attached midway up the platform could resist the largest hurricane loads. The Lena platform was installed in 1000 ft (305 m) of water. Two more "compliant" towers were installed in the Gulf of Mexico in 1998: Amerada Hess Baldpate in 1648 ft (502 m) and Chevron Texaco Petronius in 1754 ft (535 m). Petronius is the world's tallest free-standing structure.

Although nearly all of these platforms are of steel construction, around two dozen large concrete structures have been installed in the very hostile waters of the North Sea in the 1980s and early 1990s and several others offshore Brazil, Canada and the Philippine.

Among these, the Troll A (fig.1.4) gas platform is the tallest concrete structure in existence. It was installed offshore Norway in 1996. Its total height is 1210 ft (369 m), and it contains 245,000 m^3 of concrete, equivalent to 215,000 home foundations.

Gravity structures differ from other fixed structures in that they are held in place strictly by the weight contained in their base structures. The Troll platform, for example, penetrates 118 ft (36 m) into the seabed under its own weight.

▲ **Figure 1.4** Troll A gas platform – world's tallest concrete structure

The first floating production system, a converted semi-submersible, was installed on the Argyle field by Hamilton in the UK North Sea in 1975. The first ship-shaped floating production and storage system was installed in 1977 by Shell International for the Castellon field, offshore Spain. There were 40 semi-submersible floating production systems (FPSs) and 91 ship-shaped floating production and storage systems (FPSOs) in operation or under construction for deepwaters as of 2002. Petrobras has been a pioneer in pushing floating production to increasingly deeper waters in their Campos Basin fields, offshore Brazil. Table 1.1 lists the progression of field development offshore Brazil in ever-increasing water depths. Some of the unique features of innovation and records are included in the last column.

Table 1.1 Field development in offshore Brazil

Field	Well	Water Depth ft (m)	Year	Remarks
Marimba'	RJS-284-D	1255(413)	1987	Wet Christmas tree
Marlim	MRS-3	2365(721)	1991	Monobuoy & FPS
Marlim	MRL-4	3369(1027)	1994	Subsea completion
Marlin Sul	MLS-3	5607(1709)	1997	Deepest moored production unit
Roncador	RJS-436	6079(1853)	1998	FPSO depth record
2000 BC 200 Block	건-543	9111(2778)	2000	Drilling depth record at that time

Reference

Subrata Chakrabarti, Handbook of Offshore Engineering, vol.1, pp.1–10, Elsvier, 2003

Home Work Problem

1. Find the meaning of the following words or phrases by consulting the dictionary.

offshore	inaccessible	exploit	depletion
onshore	exploration	initiate	unique
installation	in detail	seafloor	seabed
moor	thruster	barge	tug
harness	explicit	focus on	offloading
hydrocarbon	to a lesser extent	date back	derrick
innovative	hostile	leap	be made up of
illustrate	indicative	guy	truss
compliant	equivalent to	penetrate	convert
submersible	ever-increasing	virtually	subsea
wellhead	shutoff	submerge	workover
abandon	result in	choke	threshold
cut-off	motivate		

2. Put T or F for the following sentences whether these are true or false, respectively, according to the facts in the passage of this unit.

- Offshore structures are fixed to the seabed.
- The offshore exploration of oil and gas began in nineteenth century.
- Nowadays deep water is defined to cover the water depth greater than 1000 ft.
- About thirty years ago, the time gap between the first drilling in 2000 ft of water and the first production was 10 years.
- The fire at the Piper Alpha oil and gas platform cost sixty-seven lives.
- COGNAC was made up of three separate structures.
- Floating structures become more and more expensive as the water depths increase compared with the fixed structures..

3. Answer the following questions.

- What is the FPSO? Please explain.
- What is the FPS? Please explain.
- Please explain the name "COGNAC".
- How much percentage of the world's oil came from deepwater in 2003?
- How much percentage of the US's oil consumption came from deepwater in 2001?
- What is the Troll A? Please explain.

UNIT 2

DEEPWATER PRODUCTION CONCEPTS

2.1 Selection of Deepwater Production Concepts

2.2 Deepwater Challenges

2.3 Offshore Disasters

2.1 Selection of Deepwater Production Concepts

The types of production concepts available for deepwater production are illustrated in figure 2.1.

Most floating production systems, and virtually all of the semi-submersible, FPSs and FPSOs, produce oil gas from wells on the seabed, called "subsea wells". Unlike wells on fixed platforms and on land, subsea wells do not allow operators to have direct access to the wells for maintenance, or for re-completion (drilling into new reservoirs from an existing well).

The well consists of a wellhead, which supports the well casing in the ground, and a pod, which contains valves to control the flow and to shutoff the flow in the case of an emergency or a leak in the riser. This pod is called a "submerged christmas tree", or simply a "wet tree". Subsea wells are expensive, but not as expensive in deepwater as placing a platform at the site. If a subsea well ceases to produce, or if its rate of production falls below economic limits, it is necessary to bring in a mobile drilling unit to remove the tree and perform the workover.

Deepwater System Types

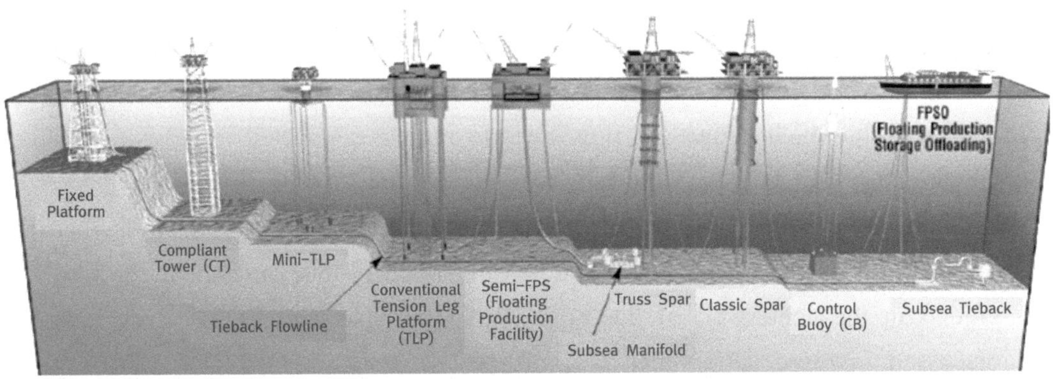

▲ **Figure 2.1** Deepwater system types [Offshore Magazine, 2002]

This can be an extremely expensive operation and if the outcome of the workover is in doubt, the operator may choose to abandon the well instead. Because of this, much of the oil and gas in reservoirs produced through subsea trees may be left behind. Subsea wells may also result in lower reservoir recovery simply because of the physics of their operation. The chokes and valves placed in a subsea tree result in a pressure drop in the flow of oil or gas. When the well formation drops below a certain threshold, production ceases to flow. The difference in cut-off pressure between a subsea well and a surface well can be as much as 1000 psi vs. 100 psi.

These facts motivated operators to seek floating platforms, which could support christmas trees at the surface, “dry trees”. Fixed and compliant platforms were safe for this kind of production because they could protect the well casings from the environment. Floating platforms generally had too much motion to protect the wells during extreme storms. A group of engineers in California invented a floating system in the early 1970s, which could be tethered to the sea floor, effectively making it a tethered compliant platform. This gave rise to what is called the Tension Leg Platform (TLP). The first commercial application of this technology, and the first dry tree completion from a floating platform, was the Conoco Hutton TLP installed in the UK sector of the North Sea in 1984. Dry trees are possible on a TLP because the platform is **heave-restrained** by vertical tendons, or tethers. This restraint limits the relative motion between the risers and the hull, which allows for flow lines to remain connected in extreme weather conditions.

The deep draft Spar platform is not heave-restrained, but its motions are sufficiently benign that risers can be supported by independent buoyancy cans, which are guided in the centerwell of the Spar. Both the Spar and the TLP designs are discussed in more detail later.

Today, many deepwater fields in the Gulf of Mexico are being developed by a combination of surface and subsea wells. Operators are able to develop a number of smaller marginal fields by combining subsea production with hub facilities. There is a growing trend towards third party ownership of the floating facilities, which opens the possibilities of several operators sharing production through one facility. A consequence of this is that floaters may be designed with excess capacity for a given reservoir, in effect adding an "option cost" into the facility investment banking on future tiebacks from additional reservoirs.

Deepwater floating production systems are generally concentrated in the "Golden triangle" of the Gulf of Mexico, offshore West Africa and Brazil. As of this writing production Spars have only been installed in the Gulf of Mexico. TLPs have been installed in the Gulf of Mexico, West Africa, the North Sea and in Indonesia. FPSOs have been installed in virtually all of the offshore oil producing areas of the world with the exception of the Gulf of Mexico. Semi-submersible FPSs are prolific in the North Sea and Brazil.

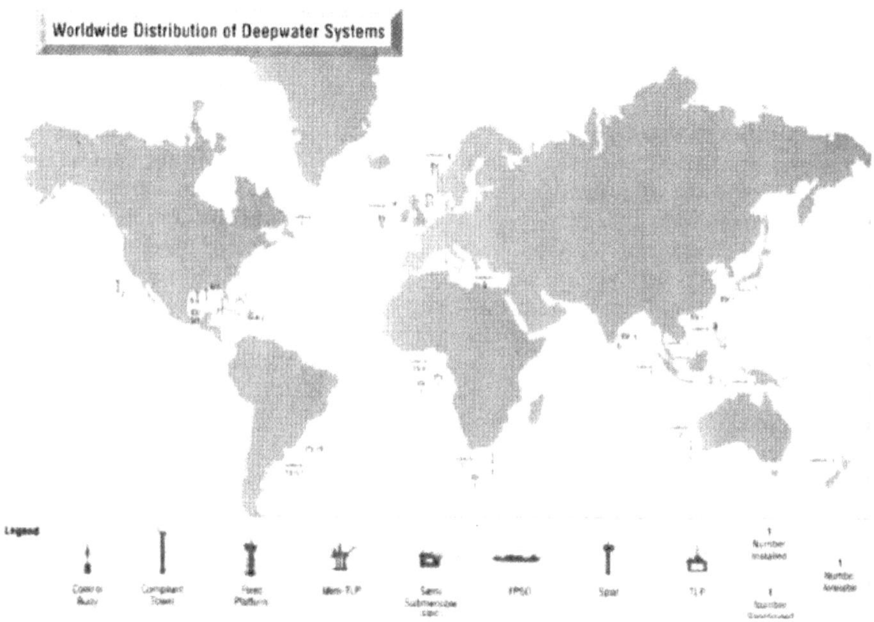

▲ **Figure 2.2** Worldwide distribution of floating production platforms [Offshore, 2002]

According to industry sources (Westwood), the floating production systems will be growing at a rate of almost 30 per year through 2006, mostly in deepwater.

There is no simple answer to the question of which concept is "right" for a particular project. Selection of a concept for deepwater production is often a multi-year effort involving numerous studies and analyses. The primary drivers are reservoir characteristics and infrastructure, which will dictate the facility size, number of wells, their location, and whether wet or dry trees are called for. Drilling often represents over 50% of the value of deepwater projects, so that the method of drilling often dictates the type of surface facilities requires, e.g. whether the facility needs to support a drilling rig or whether a leased Mobile Offshore Drilling Unit (MODU) will be used.

2.2 Deepwater Challenges

The progression of platforms placed in deeper waters worldwide through the years is illustrated in figure 2.3.

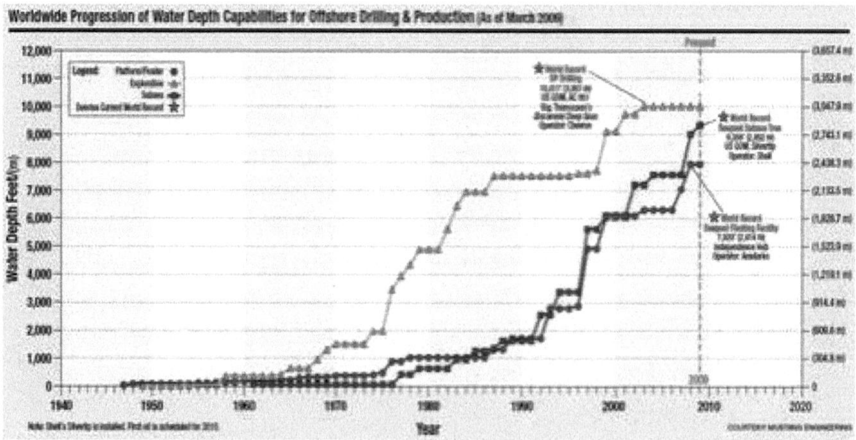

▲ **Figure 2.3** Progression of water depths [Offshore Magazine, 2002]

The figure also shows the progression of drilling and subsea completions. It is interesting to note from this figure, the gap between drilling and production. For example, the first drilling in 2000 ft of water took place in 1975. However, the first production from this water depth did not occur until 1993, i.e. 18 years later. This gap appears to be narrowing as recent advances in floating production systems and moorings have allowed rapid extension of this technology to deeper and deeper water depths.

In the sixties, production platforms designed for installation in less than a hundred meters of water were considered deep-water structures. In the seventies, platforms were installed and the pipelines were laid in nearly 300-meter water depths. FPSs and FPSO systems were designed for similar water depths in the late seventies. Early FPSs and FPSOs took advantage of a surplus supply and low cost of semi-submersible drilling units and tankers to reduce the cost of deepwater development. Currently, FPSs and FPSOs are in demand all over the world in record water depths. FPSOs are yet to operate in the Gulf of Mexico perhaps due to both regulatory requirements and the availability of infrastructure for production export, i.e. pipelines make storage unnecessary. As of this writing, the deepest floating production system is Shell's Nakika semi-submersible in 6300 ft (1920 m) water in the Gulf of Mexico. This record will be extended to 7000 ft (2133 m) in 2005 with BPs Atlantis project and there will undoubtedly be a progression into ever-deeper waters in the future.

At present, deep water is typically defined to cover the water depth greater than 1000 ft (305 m). For water depths exceeding 5000 ft (1524 m), a general term "ultra-deep water" is often used. Bottom-supported steel jackets and concrete platforms are impractical in deep water from a technical and economic point of view giving way to floating moored structures. In deep and especially ultra-deep water, risers and mooring systems provide considerable challenge. These water depths are demanding new materials and innovative concepts. Synthetic fiber ropes, which are lighter, stronger, and more cost-effective are beginning to replace wire ropes and chains. Taut synthetic polyester mooring lines produce less vertical load on the floating platform. Several deepwater floating production systems using polyester moorings are now operating in Brazil, and two Spar platforms will use polyester moorings in the Gulf of Mexico in the near future.

Flexible risers used for subsea tiebacks to floating structures are currently limited to about 5900 ft (1800 m) water depths. Steel catenary risers are becoming more common in deep and ultra-deep waters. New risers are being designed with titanium steel with high strength to weight ratio and favorable fatigue characteristics. Titanium and composite materials are also being developed for top tensioned risers.

2.3 Offshore Disasters

Although most of the offshore structures constructed to date have withstood the test time, there have been several catastrophic failures of offshore structures as well. Weather, blowout, capsizing and human errors have resulted in the loss of a substantial number of fixed and floating structures. Between 1955 and 1968, nearly two dozen mobile drilling units have been destroyed. Within the two-year period between 1957 and 1959 alone, hurricanes Hilda and Betsy inflicted losses of hundreds of millions of dollars to drilling, production and pipeline facilities. Two semi-submersibles capsized and sank in the 1980s: Alexander Keilland, an accommodation vessel in the Norwegian north sea (1980), and Ocean Ranger offshore Hibernia, Canada (1982), resulted in the loss of hundreds of lives. The worst offshore disaster occurred when the Piper Alpha oil and gas platform caught fire in 1988. One hundred and sixty-seven lives were lost. In March, 2001, the world's largest floating production system, the Petrobras P-36, sank in Campos basin (figure 2.4) costing 10 lives.

▲ **Figure 2.4** Accident of P-36 converted semi-submersible after flooding in one column [Barusco, 2002]

Reference

Subrata Chakrabarti, Handbook of Offshore Engineering, vol.1, pp.1–10, Elsvier, 2003

Home Work Problem

1. Find the meaning of the following words or phrases by consulting the dictionary.

tether	give rise to	riser	hull
heave	benign	buoyancy	hub
consequence	tieback	prolific	according to
at a rate of	infrastructure	lease	disaster
to date	catastrophic	blowout	capsize
substantial	inflict	sink	catch fire
take place	take advantage of	surplus	in demand
impractical	synthetic	taut	catenary

2. Put T or F for the following sentences whether these are true or false, respectively, according to the facts in the passage of this unit.

- Wellhead is called a "submerged christmas tree".
- Synthetic fiber ropes are lighter, stronger and more cost-effective than ropes and chains.
- The difference in cut-off pressure between a subsea well and a surface well can be as much as 1000 psi vs. 100 psi.
- Today, many deepwater fields in the Gulf of Mexico are being developed by a combination of surface and subsea wells.

- The first drilling in 2000 ft of water took place in 1995.

3. Answer the following questions.

- What is the Tension Leg Platform? Please explain.
- What is the name of the project which will be the record of the deepest flo ating production system in 2005.
- Between 1955 and 1968, how many mobile drilling units have been destroy ed?
- What is the limiting depth for use of the flexible risers used for subsea tieba cks to floating structures?

UNIT 3

OFFSHORE STRUCTURES – PART I

3.1 Introduction

The structure shown opposite is instantly recognizable as an offshore rig. This type of structure, more correctly described as a fixed steel offshore installation, or platform forms the backbone of the offshore industry and there are in excess of 7,000 such structures dotted about the oceans of the world.

Not so familiar are the structures and vessels shown elsewhere in this chapter which provide assistance to, or are in competition with the fixed steel structure. They could all be loosely described as oil rigs (or gas rigs) which gives an indication as to the ambiguity of the expression and hints at the complexity of the offshore industry.

The various rig/vessel types are:

1. Fixed steel structure.
2. Concrete gravity base structure.
3. Tension leg platform.
4. Semi-submersible vessel.
5. Floating production systems.
6. Self elevating jack-up.
7. Single point mooring.

The following is a brief description of each installation and its position within the offshore jig-saw puzzle.

3.2 Fixed Steel Structures

The traditional offshore installation shown utilizes a welded steel, tubular framework or jacket to support the topside facilities and this arrangement is referred to as a fixed steel structure. The topside facilities will vary slightly depending on whether it is oil or gas that is being produced but they will include hydrocarbon process equipment, power generation, a helideck, and accommodation and hotel services designed to cater to the needs of personnel employed in the operation and maintenance of the installation.

The single jacket installation is typical of the rigs found in deep water environments such as the northern sector of the North Sea. It should be noted that the accommodation/helideck facilities are situated as far from the potentially dangerous hydrocarbon process area as is physically possible.

▲ **Figure 3.1** Fixed steel installation

An installation may consist of any number of bridge linked jackets with modern designs tending to favor a separate jacket to house the accommodation and helideck. These multi-jacket installations tend to be restricted to shallow water developments where the construction costs are considerably less than their deep water counterparts.

Whilst the fixed steel structure is likely to remain as the industry's first choice of installation design based on operational requirements, its future use may well be restricted to the development of large fields located in intermediate water depths where a substantial return on the capital sums invested can be guaranteed. The ever changing demands of the offshore industry have spawned a number of competitors more ideally suited to specific applications.

The concrete structure has emerged as a viable alternative to the steel jacket but they both tend to suffer from the same inherent disadvantages when used to develop deep water reserves. The scantling requirements of structures that are capable of withstanding the extremes of weather conditions associated with the North Sea are considerable and this is reflected in the construction costs. Generally speaking the tension leg platform (TLP) or the floating production system (FPS) will be preferred for the development of deep water reserves. The FPS used in conjunction with sub-sea wells also represents the most cost effective solution for the exploitation of the smaller, marginal fields.

3.3 Concrete Gravity Base Structure (GBS)

Whilst the vast majority of fixed offshore platforms employ a tubular steel jacket to support the topside facilities, a number of installations have been constructed using a base manufactured from reinforced concrete. This type of installation was pioneered by the Norwegians who had experienced difficulties in producing the enormous quantities of high quality steel products and the large skilled workforce that are essential ingredients to the success of steel fabrication projects.

▲ **Figure 3.2** The Shell/Esso Brent Delta concrete gravity base structure (GBS)

The first concrete structure to be installed in the North Sea was constructed by the Norwegians in 1973 and used to develop the Ekofisk field. Since then, technological developments and improved construction techniques have lead to the use of concrete for the building of tension leg platforms, floating production systems and semi-submersible vessels.

The construction of a concrete installation base normally commences in a dry-dock. The design of the base includes void spaces or caissons suitably dimensioned to provide the structure with a natural buoyancy which will enable it to be floated clear of the dry-dock for finishing off prior to being towed to its final destination. Once on location the void spaces are flooded to facilitate positioning of the base on the sea bed whilst the topside modules are lifted into place. The void spaces may then be pumped dry and used as storage compartments for crude oil, or filled with permanent iron ore ballast. The colossal weight of the concrete structures obviates the need to install foundation piles, hence the name gravity base structure.

As to the future, the recent announcement by the Norwegian Government to exploit the Troll field using a concrete structure came as no surprise to the industry. The Troll field, which stands in 350 meters (1,150 feet) of water is currently the largest offshore oil and gas field in Europe and it will be developed using the largest concrete platform (1,270,000 tones) in the world.

Further afield, the concrete structure has been selected in preference to other designs because of its ability to withstand impact from icebergs, a feature of vital importance for installations operating in the more northerly waters of Europe, Russia and Canada. These remote locations frequently prohibit the laying of a subsea pipeline and the void spaces which feature so prominently in the design of a GBS provide valuable storage space for crude oil prior to discharge into oil tankers.

3.4 Tension Leg Platform (TLP)

The TLP concept was devised by the Conoco Oil Company as an alternative to the fixed steel structure and floating production system (FPS) for the development of deep water oil and gas fields. The world's first TLP was built by Conoco in 1984 and used to develop the Hutton field in the North Sea. Located in 148 meters (500 feet) of water, the Hutton field could quite easily have been developed using a traditional fixed steel installation but the harsh environment associated with the North Sea was judged to provide the ideal test bed for the TLP design prior to venturing into deeper waters.

The TLP fulfils a role midway between the fixed offshore installation and the floating production system. It combines the initial cost saving benefits associated with floating production systems with the operational benefits attributed to the fixed offshore installation.

The TLP consists essentially of a floating production facility which is tethered to the seabed by a number of tensioned legs. The legs are typically constructed from large diameter (250 mm to 750 mm) steel pipes or wire ropes (125 mm dia.). One end of the leg is secured in a foundation template piled into the seabed and the other end is connected to a tensioning winch within the floating superstructure. Whilst the legs remain in tension the platform can resist vertical wave induced motion.

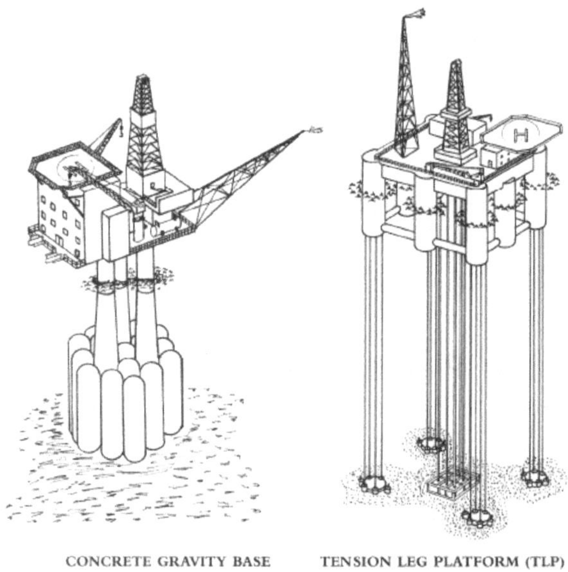

▲ **Figure 3.3** Concrete gravity base and tension leg platform(TLP)

The TLP foundation design may appear elaborate when compared to the traditional ship type anchorage arrangements employed by floating production systems but the advantages are considerable. The restriction on vertical movement permits the use of fixed installation type wellhead equipment and rigid steel riser conductor assemblies. In contrast, the floating production systems must rely on flexible riser pipes emanating from sub-sea wellheads. These incur a considerable penalty in terms of maintenance availability and costs.

The tension leg platform is still a relatively rare bird, but oil companies are beginning to appreciate the advantages offered by this unique design as they are forced to look further afield for their oil and gas. In the Gulf of Mexico, Conoco have recently commissioned their Jolliet platform in a water depth of 575 meters (1,760 feet) and Shell have chosen a TLP to develop the Auger Field (872 meters - 2,400 feet). Closer to home, Conoco have embarked on a project with the Norwegian government to construct the world's first concrete TLP which will be used to develop the Heidrun Field. Located in 325 meters (1,100 feet) of water the field is one of the largest currently under development in the North Sea and demands a structure of truly epic proportions. The four concrete foundation units which secure the Heidrun's tethers each weigh 21,000 tons, are 48 meters (160 ft) in diameter and are 26 meters (85 ft) high.

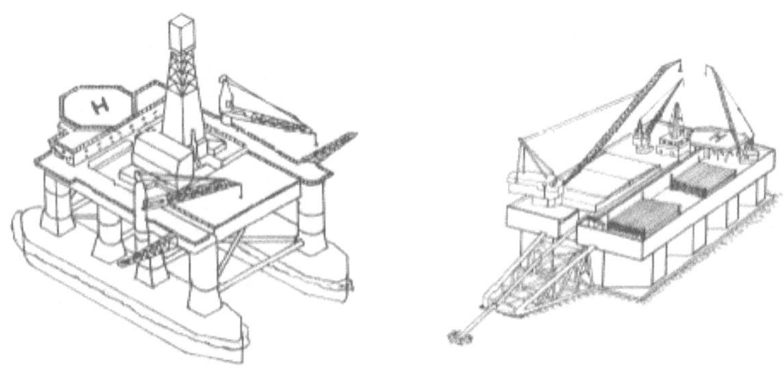

▲ **Figure 3.4** Semi-submersible drill ship and pipe lay barge

In the not too distant future we may well see further refinements of the TLP design which will permit the use of the tension legs as wellhead conductor casings thus saving on the costs of several thousand meters of steel pipe. The ingenuity of the offshore engineers appears to know no bounds.

▲ **Figure 3.5** The Shell Stadrill semi-submersible drilling vessel involved in wildcat drilling operation

Reference

Angus Mather, Offshore Engineering–An Introduction, pp.3–12, Witherby & Co., 1995

Home Work Problem

1. Find the meaning of the following words or phrases by consulting the dictionary.

in competition with		be referred to as	
ambiguity	jig-saw	helideck	intermediate
viable	alternative	inherent	scantling
in conjunction with		exploitation	marginal
crude			

2. Put T or F for the following sentences whether these are true or false, respectively, according to the facts in the passage of this unit.

- The single jacket installation is typical of the rigs found in deep water environments such as the northern sector of the North Sea.
- The concrete structure has emerged as a viable alternative to the steel jacket.
- The Troll field, which stands in 350 meters (1,150 feet) of water is currently the largest offshore oil and gas field in the world.
- The TLP consists essentially of a floating production facility which is tethered to the seabed by a number of tensioned legs.

3. Answer the following questions.

- List up the various rig/vessel types.
- When was the first concrete structure installed in the North Sea?
- The world's first TLP was built by Conoco in 1984 and used to develop the Hutton field in the North Sea. How much was the depths of water?
- Who devised the concept of TLP?

UNIT 4

OFFSHORE STRUCTURES – PART II

4.1 Semi-Submersible Vessel (SEMI_SUB)

4.2 Floating Production Systems (FPS)

4.3 Self Elevating Jack-Ups

4.4 Single Point Mooring (SPM)

4.1 Semi-Submersible Vessel (SEMI_SUB)

As can be seen from the illustrations, a semi-sub consists of an enormous box section barge supported on twin hulls. It has a colossal capacity to consume water ballast and with minimal water plane area presents an immovable object to wave action in anything but the most severe weather conditions. The design of a semi-sub does not permit the installation of an effective main propulsion plant. It is thus heavily dependent on the assistance of support ships for towage to its destination and for the deployment of anchors (typically eight). However, some vessels are fitted with computer controlled azimuthing thruster units which provide an accurate maneuvering capability, particularly when operated through a satellite navigation system. This facility enables the semi-sub to hold station in water depths in excess of 650 meters (2,200 feet) where the laying of anchors may prove impractical.

The semi-sub represents the ideal choice of vessel for performing operations where accurate station keeping and exceptional stability are prerequisite to success. Depending on the equipment fitted to the main deck the vessel may perform anyone of five roles.

i) Heavy Lift

The semi-sub will most frequently be encountered in its role as a heavy lift crane barge used for the installation of offshore platforms. Over the years the lifting capacity of the cranes has steadily increased from 2,000 tons to 7,000 tones and the more modern vessels are fitted with tandem cranes which are capable of lifts of up to 14,000 tones

ii) Accommodation

Some of the older, smaller heavy lift semi-subs are now used for modification or repair projects where the facility to provide extra accommodation and storage space take precedence over the lifting capacity of the crane. They have been joined by purpose built accommodation vessels or flotels with up to 800 beds which can remain in close attendance to fixed structures for months at a time. Transfer of personnel may be by bridge or by helicopter and again the inherent stability of the semi-sub ensures that the workforce sleep soundly, even during a winter in the North Sea.

iii) Drilling Exploration Vessel

Depending on water depth, the drilling of exploration wells will be carried out either by a self-elevating jack-up, a semi-submersible vessel or by a drill ship. The jack-up is limited to operations in water depths of approximately 120 meters (400 feet) but no such restrictions apply to the semi-sub. The current world water depth drilling record stands at a formidable 2,250 meters (7,520 feet) at a field in the Gulf of Mexico.

The combination of satellite positioning, stability and an abundance of deck space for the storage of drill pipe and well test equipment make the semi-sub the first choice of vessel for deep water exploration projects.

iv) Pipelay Barge

The sketch opposite shows a typical pipelay barge of the type which has installed nearly 10,000 kilometers of subsea pipeline in the North Sea during the past 25 years.

Once started, the laying of a subsea pipeline becomes a non-stop operation as the lay barge slowly winches its way forward on its anchors. A modern vessel can lay between 2 and 4 kilometers of pipeline a day, azimuthing thruster units often replacing the anchor spread.

Subsea pipelines vary in diameter from 3" to 42" with 30" and 36" lines being favored for the main transmission routes. The pipe is supplied to the barge in 12 meter lengths which are coated in concrete to provide corrosion protection and the negative buoyancy required to keep it on the sea bed.

On the barge the exposed ends of the pipes are welded together using a part manual, part automatic welding process. The welds are then X-rayed and if free from defects are given a protective, coating of bitumen before being launched into the sea from the stinger framework situated on the stern of the vessel. The stinger prevents buckling of the pipe as it enters the water.

The completed pipeline will be subjected to a hydrostatic pressure test to prove its structure integrity before being trenched or covered in rocks to protect it from hazards such as fishermen's nets and ships anchors.

4.2 Floating Production Systems (FPS)

A floating production system is in effect a floating oil rig. It contains all the equipment associated with a fixed installation and is used in conjunction with subsea wellheads to exploit moderate to deep water oil fields. The FPS is particularly suitable for the development of oil reserves where the installation of a fixed structure would be either impractical or prohibitively expensive.

The majority of FPS's employ a semi-submersible barge to accommodate the process plant, although more recently there has been a significant increase in the number of companies opting for monohull designs, normally employing converted oil tankers. A spread of 8 to 12 anchors ensures that they remain on location whilst producing oil, the oil being extracted from the subsea wells through flexible flowlines designed to absorb any wave induced motion which might affect the position of the vessels.

The increase in the popularity of monohull floating production systems is largely due to the fact that processed oil can be stored on board the vessel prior to export, making them in effect Floating, Production, Storage and Offloading (FPSO) facilities. The oil is eventually discharged into shuttle tankers for transportation to a refinery.

The processed oil from semi-submersible floating production systems may be discharged into a subsea pipeline, or transferred to a shuttle tanker via a single buoy mooring (SBM) for onward transportation.

The beauty of the floating production system lies in the fact that it can simply lift anchors and depart to pastures new when the oil reserves reach a commercially unprofitable level. This attractive feature has lead to the evolution of a monohull FPS which can be used to develop marginal reserves located in more moderate water depths. Pioneered by British Petroleum (BP), these designs are often referred to as PSV's (Production Storage Vessel) and a dynamic positioning capability enables them to remain on station over the wells without the need to deploy anchors. Once the storage tanks are full (typically 20 days production), the vessel disengages from the wellhead and proceeds to a shore terminal to discharge the cargo.

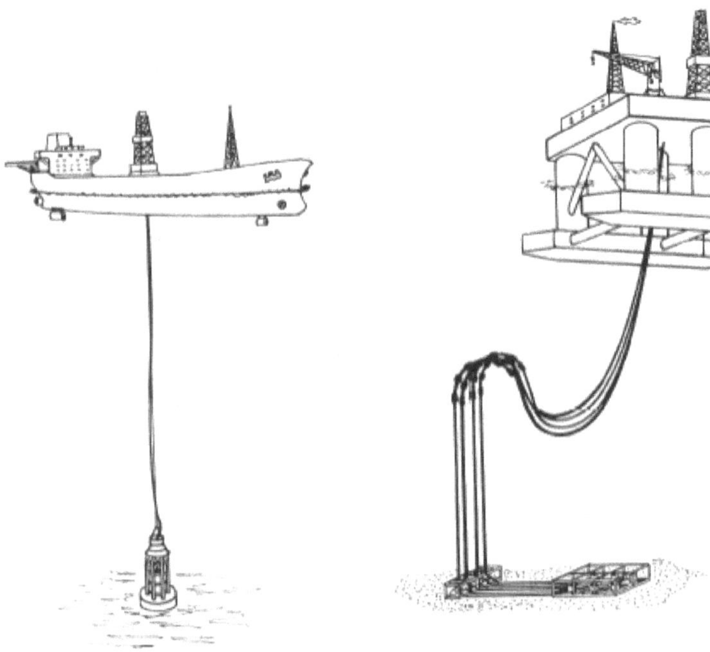

▲ **Figure 4.1** Floating Production Systems(FPS)

Current predictions indicate that floating production systems will continue to feature prominently in the future development plans of the leading oil companies. They offer considerable cost saving benefits when compared with fixed structures and with the introduction of the driverless subsea wellhead, there is virtually no limitation to the water depth in which they can operate. They are used extensively in the Campus Basin of Brazil, the world's deepest subsea development in water depths up to 1,000 meters (3,400 feet). Floating Production Systems have also been selected for the development of the new discoveries in the Atlantic Ocean, West of Shetland.

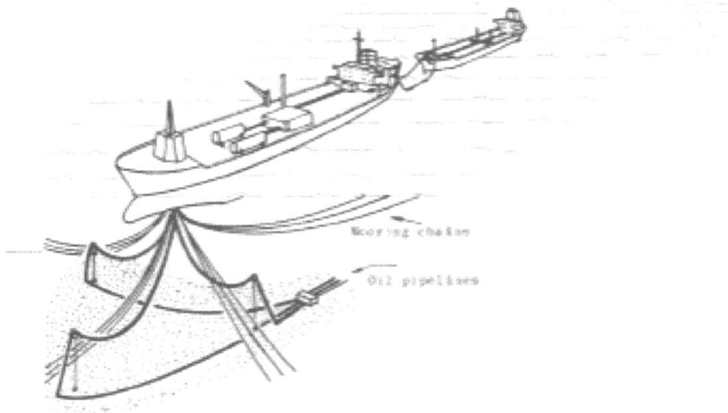

▲ **Figure 4.2** Floating Production Storage and Offloading Systems(FPSO)

4.3 Self Elevating Jack-Ups

The self elevating jack-up is one of the stalwarts of the offshore industry, its origins dating back over 60 years when it could be found searching for oil in the muddy swamps of Louisiana.

The self elevating jack-up consists of a triangular, sometimes rectangular, shaped box section barge fitted with three (sometimes four) moveable legs which enable

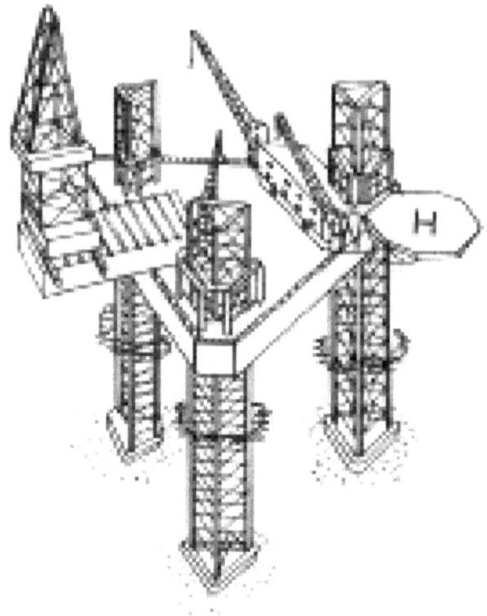

▲ **Figure 4.3** Self Elevating Jack-Ups

the vessel to stand on the sea bed in water depths of up to approximately 120 meters (400 feet). The shape of the vessel imposes design restriction similar to those encountered with semi-submersible vessels which prevent the fitting of a main propulsion unit. Transportation over short distances is effected under the tow of tugs or supply boats whilst it is both quicker and safer to accommodate a jack-up on the back of a submersible heavy lift ship to undertake long sea passages. Once on location, thruster units enable the vessel to maintain position whilst the legs are lowered to the sea bed and the hull is jacked into position, clear of wave action. The base of each leg is fitted with a "spud can" which consists of a plate or dish designed to spread the load and prevent over penetration of the leg into the sea bed. High pressure jets of water or compressed air may be used to remove loose debris in the vicinity of the spud cans whilst the legs are maneuvered into position. This process is known as "spudding in".

▲ **Figure 4.4** The jack-up Neddrill 3 testing and exploration well. The gas is being flared from a temporary overboard vent boom.

Jack-ups are primarily used for drilling operations but a number have been constructed to act specifically as accommodation support vessels to provide assistance to fixed installations during construction, modification or repair programs. The jack-ups used for drilling operations fulfill two completely different roles. The majority of jack-ups are used for exploration purposes (wildcat drilling) and they lead a solitary existence searching for oil or gas in some of the remotest corners of the globe.

The other role in which jack-ups will be encountered is in the drilling of wells for permanent installations, particularly gas producing platforms such as those located in the southern sector of the North Sea. The gas production process does not require the same level of ongoing drilling activity and well modification work that is associated with the production of oil. Consequently derricks are not normally fitted to gas producing installations and all well modifications and drilling operations must be carried out by a jack-up sited alongside. The drilling derrick is fitted to rails so that it can be cantilevered into a position which provides direct access to the wellhead area.

The future of the self elevating jack-up looks assured with old rigs being replaced with designs capable of operating in ever deepening, harsh water environments. In fact the jack-up looks set increase its share of the offshore 'market with a new generation of rigs designed to operate as process facilities for the development of marginal oil fields using subsea wells.

4.4 Single Point Mooring (SPM)

The single point mooring concept originated as a solution to the problem of transferring crude oil from an onshore reception facility or refinery into very large oil tankers which were physically too big to enter port. The tankers are moored to a large buoy located at some considerable distance from the coast, often several miles, the buoy itself being secured by a spread of anchors. The oil is transferred from a subsea pipeline, through a swivel connector in the buoy into loading hoses attached to the tanker, thc tanker thus being free to weathervane around the buoy independent of both wind and tide.

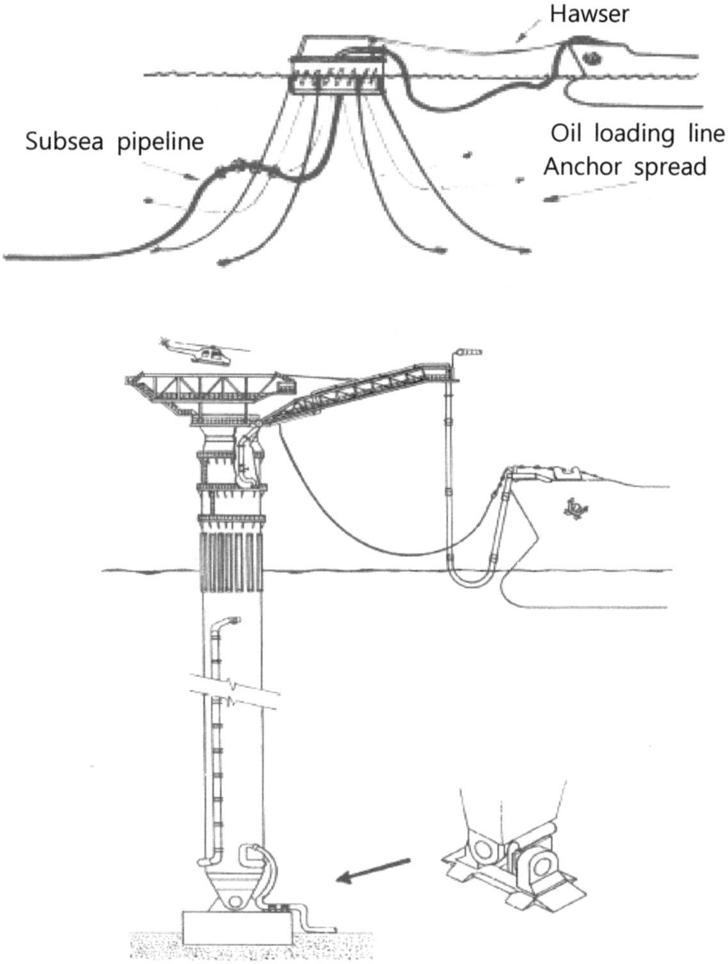

▲ **Figure 4.5** Calm buoy and single point mooring

Single point moorings of the type just described employing a tethered buoy and flexible riser are still used extensively and a number of more sophisticated designs such as the CALM (catenary anchor leg mooring) yoke and the FRAMS (flexible riser and mooring systems) have emerged to service the more specialist needs of the offshore industry. Variations on the single point mooring theme have also been developed which employ a rigid structure or riser column to convey crude oil from the sea bed to the surface. Examples of this particular breed are the SALM (single anchor leg mooring) and the ALP (articulated loading platform). Regardless of design they all attempt to achieve the same end result which is to provide a safe mooring for the loading vessel in wind speeds of 80km/hour (50 mph) and wave heights of 5.5 meters (18 feet).

A brief account of a fixed column SPM will now be given which is representative of the type used as a loading terminal by Statoil for their Statfjord B development.

This particular SPM consists of three main components:

i) Gravity Base

The gravity base shown measures 20 x 20 x 8 meters (65 x 65 x 27 feet), weighs 950 tons and contains 4,500 tons of iron ore ballast to assist the forces of gravity in keeping the column firmly rooted to the sea bed.

ii) Column

The steel column boasts dimensions equally as impressive as those of the gravity base measuring 170 meters (560 feet) in height, 9 meters (30 feet) in diameter and weighing 4,000 tons. To assist emplacement the column is filled with 2,400 tons of iron ore in addition to the permanent water ballast. A Cardan type articulated joint provides the means of attachment to the base a feature that also permits a degree of lateral movement. The crude oil is transferred from the manifold at the gravity base up to the rotating head by two 36 inch diameter (750 mm) pipelines routed up the outside of the column.

iii) Rotating Head

The rotating head contains the quick release mooring and loading attachments, temporary accommodation and a helideck and is designed to act as a moving pivot for a ship to rotate around under the influence of wind and tide. The oil is transferred from the column into the loading hoses through a swivel coupling at discharge rates of up to 57,000 barrels an hour (8,000 tones) and this particular SPM has the facility to load ships ranging in size from 80,000 to 150,000 tones.

Riser column SPM's such as those just described have two main applications. Connected to a subsea pipeline the riser towers provided the ideal means of loading tankers, for whilst a pipeline is certainly the most cost effective means of conveying large quantities of crude oil over long distances the SPM enables tankers to load oil destined for global export without the inconvenience and expense of entering port. The other area where the SPM has grown in popularity in recent years is the subsea market. When a field is developed exclusively by subsea wells, the oil is normally processed by a floating production system (FPS) prior to export. The wells may be connected to the floating production system by individual flexible risers, or manifolded on the sea bed and connected by an SPM, an altogether much neater arrangement.

Reference

Angus Mather, Offshore Engineering–An Introduction, pp.3–12, Witherby & Co., 1995

Home Work Problem

1. Find the meaning of the following words or phrases by consulting the dictionary.

colossal	obviate	afield	exclusively
iceberg	northerly	prominent	devise
attribute	template	elaborate	anchorage
emanate	incur	embark	epic
towage	prerequisite	tandem	formidable
abundance	bitumen	stinger	trench
in effect	prohibitively	refinery	onward
pasture	disengage	stalwart	swamp
swivel	yoke	articulate	emplacement
ore	lateral		

2. Put T or F for the following sentences whether these are true or false, respectively, according to the facts in the passage of this unit.

- A semi-sub consists of an enormous box section barge supported on twin hulls.
- The semi-sub represents the ideal choice of vessel for performing operations where accurate station keeping and exceptional stability are not so important.

- The current world water depth drilling record stands at a formidable 1,500 meters at a field in the Gulf of Mexico.
- The single point mooring concept originated as a solution to the problem of transferring crude oil from an onshore reception facility or refinery into very large oil tankers which were physically too big to enter port.
- Jack–ups are primarily used for drilling operations.

3. Answer the following questions.

- Semi–submersible vessel may have five functions. What are they? List at least three of them.
- Explain the floating production system in detail.
- What is self elevating jack–ups? From where did it originate?

UNIT 5

OFFSHORE CONSTRUCTION EQUIPMENT --BARGES

An offshore construction barge must be long enough to have minimal pitch and surge response to the waves in which it normally works, wide enough in beam to have minimum roll, and deep enough to have adequate bending strength against hog, sag, and torsion, as well as adequate freeboard. The deck plating must be sufficiently continuous to enable it to resist the membrane compression, tension, and torsion introduced by wave loading.

Impact loadings can come from wave slam on the bow, from ice, and from boats and other barges hitting against thc sides. Unequal loads may be incurred in bending of the bottom hull plates during intentional or accidental grounding and of the deck plated due to cargo loads. Corrosion may reduce the thickness of hull plates.

The internal structure of a barge is subdivided by longitudinal and transverse bulkheads. Because of the relatively high possibility of rupturing of a side plate, with consequent flooding of the adjacent compartment, the longitudinal bulkheads are usually spaced at the middle third of the beam. A single centerline bulkhead could allow flooding of one entire side, causing excessive heel and possible capsizing.

Longitudinal bulkheads plus the two sides provide the longitudinal shear strength of the barge. The transverse bulkheads are usually spaced with one just aft of the bow (the collision bulkhead), one forward of the stern, and one or more in the midships region. These provide the transverse shear strength. Quartering waves produce torsion as well as bending in both planes. The torsional shear runs around the girth of the vessel sides, deck, and bottom.

Typical offshore barges run from 80 to 160 m in length. Width should be one third to one fifth the length. Depth will typically run from 1/12 to 1/15 the length. Such ratios have been found to give a reasonable balanced structural performance under wave loadings. Inland barges, subjected to minimal wave loadings and required for operations in shallow water may have depths as low as 1/20 of the length. They may be stiffened by external trussing. Shallow-depth barges are often used in rivers and lakes; these can be very hazardous in exposed locations where not only high quasi-static bending stresses can develop but also dynamic amplification and resonance.

Offshore barges typically have natural periods in roll of 5 to 7 s. This is unfortunately the typical period of wind waves; hence resonant response does occur. Fortunately, damping is very high, so that while motion in a beam sea will be significant, it reaches a situation of dynamic stability. The corners of barges are subject to heavy impacts during operations; thus they must be heavily reinforced. Fenders should be provided on the corners to minimize impact damage to other craft and structures. Fenders should be provided along the sides to minimize damage to the barge itself from other boats and barges as they are docked. These may be a combination of integral fender strakes plus renewable fenders. Bitts are provided at the corners and at intervals along the sides to enable the securing of the barge and any other craft which come alongside. Towing bitts are provided on both bow and stern.

Consideration must be given to the need to temporarily weld padeyes to the deck in order to secure cargo for sea.. These padeyes must distribute their load into the hull; they cannot develop proper strength by just welding to the deck plate. They will be subjected to impact loads in both tension and shear. In modern offshore barge design, special doubler plates are often affixed over the internal bulkheads so that padeyes may be attached along them. Low-hydrogen electrodes should be used. Alternatively, posts may be installed, running through the deck to be welded in shear to the internal bulkheads.

The deck is often protected by timbers to absorb the local impact and abrasion of the load. This is especially needed for barges which will carry rock which will be removed by clamshell or dragline bucket, or upon which a tracked crane or loader will operate. Manholes are provided in the deck for access to the inner compartments. These must be watertight. There should be a heavy coaming to protect the dogs or bolts which secure the manhole. Once again the warning must be made about entering inner compartments which have been closed for a long

▲ **Figure 5.1** Steel tubular piles being transported from Korea to Bangladesh by barge. Note structural supports at sides.

period. They are probably devoid of oxygen and must be thoroughly aerated before entry.

Marine barges are often intentionally grounded (beached) in order to load or unload cargo. The beaching areas must be well leveled and all boulders and even large cobbles removed in order to avoid holing or severely denting the barge. Once beached, the barge should be ballasted down so that it will not be subjected to repeated raising and banging down under wave action at high tide.

When heavy loads are skidded on or off a barge, they punish the deck edge and side because of the concentrated loading. Skid beams are often arranged to partially distribute the load to interior bulkheads. A timber "softener" may be temporarily bolted to the deck edge. The barge must be analyzed structurally for each stage of loading to ensure that a side or bulkhead will not buckle under the temporary overload.

Cargo must be secured against movement under the action of the sea. See figure 5.1. Thus sea fastenings are designed to resist the static and dynamic forces developed under any combination of the six fundamental barge motions: roll, pitch, heave, yaw, sway, or surge. The dynamic component is due to the inertial forces which develop due to acceleration as the direction of motion changes. Roll accelerations are directly proportional to the transverse stiffness of the barge, which is measured by its metacentric height. Since barges typically have large metacentric heights, accelerations are severe. Conversely, if due to high cargo, the metacentric height is low, the period and amplitude of roll and the quasi-static force imparted by the load are greater, but the dynamic component may be less.

These loads are cyclic. Sea fastenings tend to work loose. Wire rope stretches; wedges and blocking fall out. Under repeated loads, fatigue may occur, especially at welds. Welds made at sea may be especially vulnerable because the surfaces may be wet or cold. Low-hydrogen electrodes will help. Chains are a preferred method for securing cargo for sea, since chain does not stretch. If structural posts are used, they should be run through the deck to be welded in shear to the internal bulkheads. The slot through the deck should then be seal-welded to prevent water in-leakage.

The effect of the accelerations is to increase the lateral loading exerted by the cargo due to the inclination of the barge by a factor of two or more. Flexing of the barge can also have a significant effect on support forces and the sea fastenings. Thus deeper and hence stiffer barges will experience a smaller range of loads than shallow, less stiff barges.

With important and valuable loads such as modules or jackets, sufficient freeboard should be provided to ensure stability even if one side compartment or end compartment of the barge has been flooded.

Barges are normally designed to the standard loading criteria of the classification society. These criteria are usually based on submergence of the hull to the deck line, plus an arbitrary load of 3 m of water on deck. The typical barge is not designed for complete submergence.

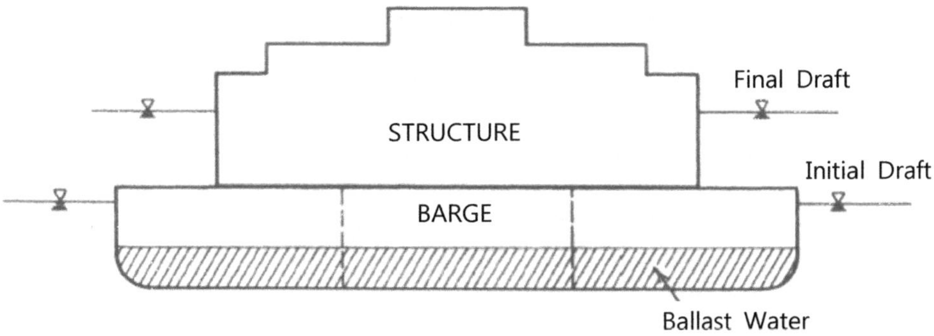

▲ **Figure 5.2** Launching a buoyant structure from a construction barge.

Proposals are often made to build a structure on a barge, then submerge the barge by ballasting, and float the new structure off. Actually this has been successfully carried out in a number of cases: the construction of the pontoons of a dry-dock in northern Spain, the manufacture of several hundred shallow draft concrete hulls and posted barges near New Oleans, and the construction of Arctic offshore caissons in Japan. See figure 5.2. However, there are three key items that have to be taken care of:

1. When a barge is submerged by partially filling compartments with ballast, the external pressures are essentially the same as if the barge were empty and submerged to that depth. The hull must be designed for the deepest submergence.

2. Once the deck of the barge is submerged, stability of the barge itself is lost, although the structure on deck may give an effective water plane and thus provide stability during submergence. Once the structure floats free, while the barge itself may initially have stability due to the low center of gravity of the ballast water, as it is pumped, the barge may rise up in an unstable mode, out of control.

Stability and control can be provided by columns at the ends which always extend above the waterline. See figure 5.3.

3. The third problem is that of depth control. The support barge must be neutrally or negatively buoyant in order to launch the structure. Control of depth can then only be achieved by:

   a. Carrying the operation out in shallow water of known depth;
   b. Use of lowering lines from surface barges; however, the load distribution may shift, causing lines to fail successively - even small waves may develop large forces in lines due to the heave of the lowering vessels;
   c. Use of columns penetrating the waterline, as recommended for stability control; these will tend to regulate the load distribution.

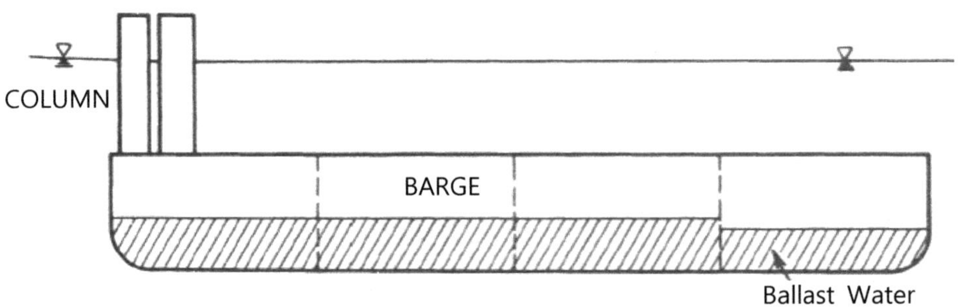

(a)

(b)

▲ **Figure 5.3** (a) Maintaining stability and draft control by use of columns at one end. (b) Semi-submersible vessel for long-distance transport of dredges, crane barges, and caissons. (Courtesy of Chevron Resources Canada Ltd.)

If a barge is to be seated, fully submerged, on an underwater embankment or on the seafloor in relatively shallow water, then it can be tipped down, one end first. Thus the beam of the barge and the inclined water plane provide stability at this stage. Then the barge end touches bottom. Now the barge may be fully submerged, gaining its stability from the end of the barge reacting against the bottom. See figure 5.4. This practice is normally limited to a water depth about one third that of the barge length.

Note that as the barge is tipped down, the transverse water plane area and moment of inertia is reduced to about one half of normal. Therefore, transverse instability can develop before the end touches the bottom, and the barge can roll. This acts to limit the depth of water suitable for such an operation. To recover the barge from the seafloor, the reverse procedure is followed, raising one end first.

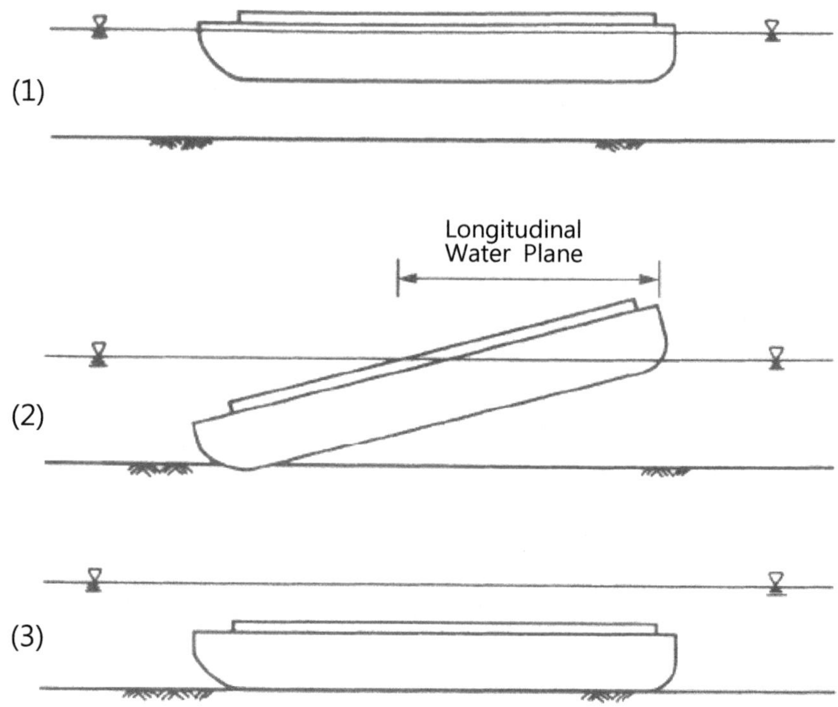

▲ **Figure 5.4** Tipping a barge to the seafloor in shallow water.

A barge seated on a mud or clay seafloor develops a suction effect, consisting both of adhesion and a true suction due to differential water pressures. To break the barge loose requires that full hydrostatic water pressure be introduced under the bottom and that the adhesion of the clay to the barge be broken.

Extensive experiments by the Naval Civil Engineering Laboratory at Port Hueneme, California, and confirmed by practical experience in the Gulf of Mexico show that breaking loose can best be accomplished by sustained water flooding at a low pressure, less than the shear strength of the clay. Higher pressures will just create a piping through to the sea and prevent development of any pressure. Periods of several hours may be required to develop a fully equalized head of water under the structure. Then by deballasting one end first, the barge can be lifted clear. This method was used to free the GBS-1 (Super-CIDS) exploratory drilling platform, after 1 year's service seated on a clay foundation. It proved very successful.

As with ship salvage, a fully submerged barge must be given only limited positive buoyancy; otherwise it may break loose suddenly. If compressed air has been used to displace water in open compartments, the vessel will achieve additional buoyancy with every meter of rise due to the expansion of the air inside and may become uncontrollable. For all the above reasons, submergence of standard barges must be considered only to shallow depths. For deeper submergence, special construction and internal pressurization may be required; these are described in a later chapter.

Reference

Ben C. Gerwick, Jr., Construction of Marine and Offshore Structures, 2ndEd.,pp.102–108,CRCpress,2000

Home Work Problem

1. Find the meaning of the following words or phrases by consulting the dictionary.

surge	hog	sag	torsion
bow	longitudinal	transverse	bulkhead
heel	transverse	stern	shear
girth	shallow	quasi-static	amplification
resonance	reinforce	fender	strake
bitt	padeye	clamshell	coaming
devoid	aerate	boulder	cobble
fastening	inertial	be proportional to	metacentric
vulnerable	submergence	pontoon	caisson
embankment	adhesion	salvage	

2. Put T or F for the following sentences whether these are true or false, respectively, according to the facts in the passage of this unit.

- The internal structure of a barge is subdivided by longitudinal and transverse bulkheads.
- Typical offshore barges run from 50 to 100 m in length.
- Offshore barges typically have natural periods in roll of 2 to 3 s.
- The deck is often protected by steels to absorb the local impact and

abrasion of the load.

- Once the deck of the barge is submerged, stability of the barge itself is lost, although the structure on deck may give an effective water plane and thus provide stability during submergence.

3. Answer the following questions.

- According to the passage in this unit, how much natural period do typical offshore barges have?
- Explain three methods of depths control of barges.

UNIT 6

TOPSIDES FACILITIES DESIGN - PART I

6.1 Introduction

As described previously, the general arrangement and layout of ship-shaped offshore units designed for oil and gas operations may be grouped into several major parts: hull structures including storage tanks, topsides (processing facilities), export facilities, mooring facilities, accommodations, machinery space, subsea systems, and flowlines. All of these various parts are equally important to achieve successful operation, with due consideration of safety, health, environment, and costs versus benefits.

This chapter focuses on topsides, moorings, and export facilities. The material presented herein is aimed at the nonspecialist introductory reader. It is consistent with the content of this book and is included, primarily, to complete the coverage of the various aspects relating to ship-shaped offshore units.

Topsides consist of processing facilities that are typically located as elevated modules that are several meters (say, 3m or more) above the main deck of the vessel hull, but related piping systems may be located on the main deck of the vessel hull. Depending on the vessel size and topsides layout, the topsides modules may have multiple decks that contain the oil-, water-, and gas-processing facilities; utility systems; and similar functions. The preferred configuration, however, may be that to the extent possible, the topsides facilities would be incorporated as single-layer "pancake" units. The single-layer unit arrangement requires a larger main deck area for a given set of needs.

Offloading or export facilities are used to unload cargo to shuttle tankers that will be appropriately moored, say, at buoys, and may also possibly use dynamic position systems to keep the tanker in place during offloading. For ship-shaped offshore installation itself, it is important that the installations have its own mooring facilities to achieve the required station-keeping of the vessel both during production and offloading. Either a weathervaning turret system or a spread-mooring system will be primarily employed for this purpose.

Generally, the design and construction of topsides modules and mooring systems are carried out by independent specialized contractors in parallel with the hull design construction, although a single engineering procurement and construction (EPC) contractor may manage the entire project process until commissioning. The topsides modules are typically prepared as preassembled units (PAU) that are then mounted onto the deck of the floating offshore unit using heavy-lift cranes. The various complexities essentially mean that related structural design issues pertaining to the interface between hull and topsides modules, including modules supports, must be considered in a great detail during design, construction, and installation. For example, there are a number of design teams involved for the various disciplines, and their designs are interdependent in reality. Therefore, such interdependent design data must be transferred back and forth between interfaces, and among disciplines such as topsides and mooring, at appropriate points in time. Related data accuracy, consistency, and timeliness must be maintained, and the sequence and progress work of the various disciplines involved must be appropriately monitored and controlled.

This chapter presents some design considerations for topsides, mooring, and export facilities in terms of functional requirements and also considering their structural interfaces to the vessel hull.

6.2 Topsides Facilities

The topsides facilities may consist of several types of modules, skids, units, or facilities, each addressing one of the following functions:

- Oil and water separation
- Gas compression

- Water injection
- Cargo handling and offloading
- Utility and support
- Safeguards

▲ **Figure 6.1** (a) A bird's eye view of an FPSO with an external turret-mooring system(Courtesy of Samsung Heavy Industries). (b) A Computer graphic of the topsides facilities including flare tower(Courtesy of Samsung Heavy Industries)

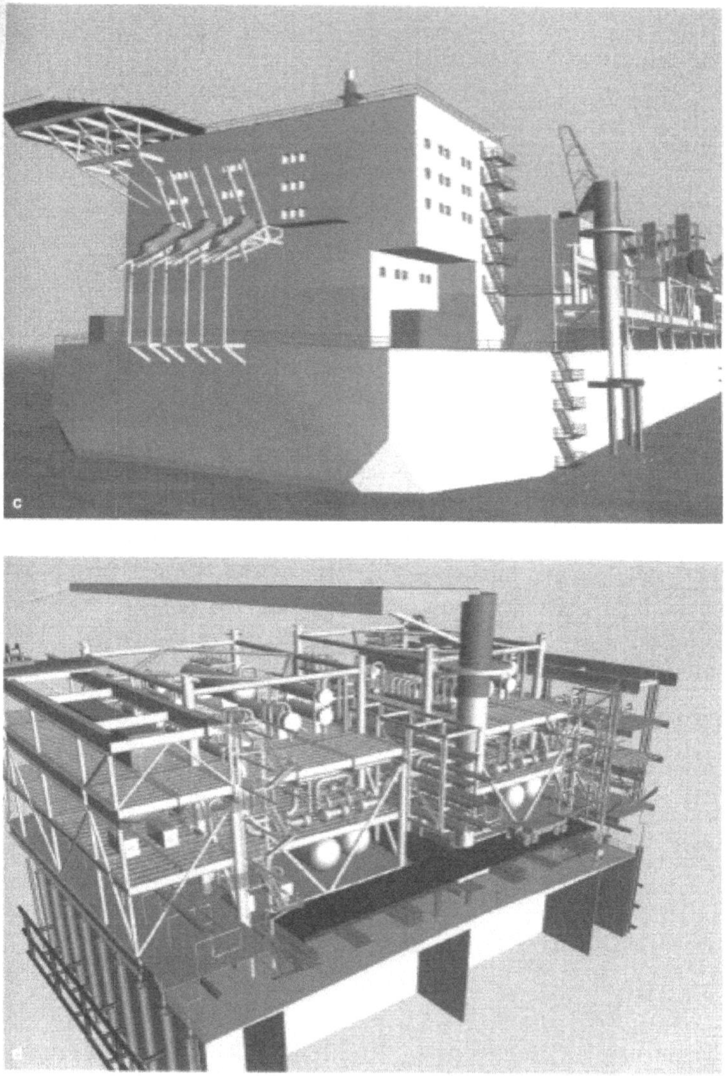

▲ **Figure 6.1 (cont)** (c) A computer graphic sea-level view of the accommodation area at stern with fire water pumps at both sides and free-fall lifeboats(Courtesy of Samsung Heavy Industries). (d) A computer graphic view of crude separation modules equipped with maintenance lanes for removal of equipment(Courtesy of Samsung Heavy Industries).

▲ Figure 6.1 (cont) (e) A computer graphic view of central pipe-rack(Courtesy of Samsung Heavy Industries). (f) A computer graphic view of ladders for escape from topsides level to lower main deck level(Courtesy of Samsung Heavy Industries).

Figure 6.1 shows example design graphics of various topsides modules in an FPSO. For the design of topsides facilities, as well as other structures or facilities of ship-shaped offshore installations, risk management and control in terms of safety, health, the environment, and financial expenditures are necessary. Active and passive fire protection measures must be installed, including fire and blast walls. Strategies and means for escape, evacuation, and rescue must be considered, developed, and implemented.

In most jurisdictions, an environmental impact study is invariably required to ensure that the impacts associated with the operation of the offshore installation are minimal in terms of the interactions between the offshore installation (including the export system and subsea infrastructure) and the native ecology of the field (e.g., fish breeding grounds, cetacean migration routes, seabird activity, and seabed flora/fauna such as corals).

In the following sections, a general description of the functions, components, and selected design considerations for important items of topsides facilities in an offshore unit meant for oil and gas production, storage, and offloading are discussed. Note that the following discussion is not meant to be comprehensive. The systems involved are not unique, as many variations can be found in practice. Related and more comprehensive discussions as well as design guidelines for topsides facilities are presented in Lapidaire and de Leeuw (1996) and UKOOA (2002).

6.2.1 Oil and Water Separation Facilities

Crude oil is stabilized and dewatered by the oil and water separation facilities. Typically, multiple stages are required for the processing of segregation. The stabilized crude oil may typically flow through crude coolers to a wet crude reception tank, which is usually a center cargo tank of the vessel. In there, the oil is further separated by gravity settling to basic sediment and water. The stabilized dry crude oil that comes out of the reception tank is then sent to the cargo oil tanks for temporary storage before export to shore by shuttle tankers.

The produced water may be disposed overboard after deoiling in hydrocyclones and degassing in a produced water flash drum. The reject streams developing in hydrocyclones will be sent to the low-pressure flare drum. Gas coming out of the produced water flash drum will also be sent to the low-pressure flare system.

For possible discharge overboard, the oil-in-water content must be sufficiently low to prevent marine pollution, say 40 parts per million (ppm) (DTI 1971). Recent trends show that the produced water is more often reinjected into the reservoir not only to provide pressure support but also to prevent potential marine pollution possibly arising from continuous overboard disposal.

With the reinjection option, the produced water may be injected on its own or with treated seawater. The seawater treatment is commonly carried out by the use of batteries of hydrocyclones; such operations can sometimes result in low levels of changes to the ambient hull motions.

Consideration must be given to handling, storage, and disposal of the produced water in emergency cases when the injection pumps may fail to work. For outages of short duration, the produced water may be stored in a separate tank. In the event that the pump is not available for a longer period, the produced water may be disposed overboard, provided it meets the necessary oil content specification, say 40 ppm.

6.2.2 Gas Compression Facilities

The gas compression facilities that include gas coolers and suction scrubbers compress the gas by electrical motor-controlled compressors; these must usually operate at a constant speed with constant discharge pressure. Typically, multiple processing stages of gas compression are necessary to keep the pressure above a specified level. In the early years of field production, the reservoir gas pressure may be relatively high. In this case, gas coming out of the processing stages during degassing may sometimes be mixed with the gas from the high-pressure gas compressor.

In the later years of production, the reservoir gas pressure will usually have decreased compared to the early years. In this case, gas from the separation stages may be sent to the inlet of the high-pressure gas compressor for pressure increase by additional compression as required.

The gas is dehydrated in the glycol contactor and then compressed to the export pressure in the export gas compressor. The glycol contactor is an absorption column where the gas is contacted with tri-ethylene glycol to remove water vapor from the gas. The water-rich tri-ethylene glycol is then sent from the contactor to be regenerated in the tri-ethylene glycol regeneration package. The gas will be exported through a gas line or other means with the appropriate fiscal metering provided by a gas-metering package.

6.2.3 Water Injection Facilities

For reservoir support during production in the field, high-pressure water usually needs to be injected to the reservoir. The water to be injected will be filtered or deaerated seawater or produced water or perhaps a mixture of them. Some additional treatment may also be carried out by including addition of corrosion inhibitors, scale inhibitors, or biocides to the water.

Oxygen will be removed from seawater by vacuum deaeration or by nitrogen stripping. Seawater filtration will be performed by coarse filters first and then by fine filters. The filter unit selection will consider line-cleaning requirements.

The water injection will be carried out via high-pressure centrifugal pumps. Only a single high-pressure pump may need to be used in some cases when the required water volume and the water injection criticality for reservoir support are limited, but multiple pumps may be required in the more demanding cases. In the design of high-pressure injection pipe work, the possibility of high-surge pressures arising, for example, from the rapid closure of valves on the subsea injection flow lines will need to be considered.

Reference

J.K.Paik & A.K.Thayamballi, Ship-shaped Offshore Installations, pp.318-327, Cambridge University Press, 2007

Home Work Problem

1. Find the meaning of the following words or phrases by consulting the dictionary.

layout	offload	contractor	pertain
interdependent	discipline	expenditure	jurisdiction
ecology	cetacean	migration	flora
fauna	corals	hydrocyclone	degas
flare	outage	scrubber	contactor

2. Put T or F for the following sentences whether these are true or false, respectively, according to the facts in the passage of this unit.

- Offloading or export facilities are used to unload cargo to shuttle tankers that will be appropriately moored.
- The design and construction of topsides modules and mooring systems should not be carried out by independent specialized contractors in parallel with the hull design construction.
- Typically, only single stage is required for the processing of segregation.
- For possible discharge overboard, the oil-in-water content must be sufficiently low to prevent marine pollution, say 400 parts per million (ppm)
- Typically, multiple processing stages of gas compression are not necessary to keep the pressure above a specified level.

3. Answer the following questions.

- The topsides facilities may consist of several types of modules, skids, units, or facilities. What are their functions?
- Where is the gas compressed to the export pressure?.
- What are the general arrangement and layout of ship–shaped offshore units designed for oil and gas operations?. Please group into several major parts.

UNIT 7

TOPSIDES FACILITIES DESIGN – PART Ⅱ

7.1 Cargo Handling Systems

7.2 Utility and Support Systems

7.3 Safeguard Systems

7.1 Cargo Handling Systems

A well-through-out cargo management philosophy is necessary for the efficient and safe handling of the oil cargo. Some of the design issues that will need to be addressed may relate to the following:

- Offloading to shuttle tankers
- Reception tanks
- Transfer between tanks and associated valving
- Inerting and venting of tanks
- Drainage
- Tank inspection and isolation for entry
- Tank washing
- Cargo pump type and maintenance of cargo pumps
- Protection against over- and under-pressure in tanks
- Manifold configuration and isolation
- Product heating and cooling
- Backups in case of equipment breakdown
- Safety and accidents (e.g., leaks, pollution, fire, explosions)

Offloading of cargo to shuttle tankers is considered in Section 7.5. A reception tank is needed on board to store the stabilized crude oil that will be further dehydrated by gravity, settling to a specified basic sediment and water content, as described in Section 6.2.1. Transfer of oil to a shuttle tanker is carried out via a transfer hose as described in Section 7.5, and the safety of valve systems must be assured to reduce the risk of tank overpressure, for example.

For the safe handling and discharging of cargoes or other fluids, inerting of the spaces involved is to be carried out. Two inert gas systems are usually pertinent for ship-shaped offshore installations, one for blanketing and purging of production equipment and the other for blanketing of main cargo tanks. The system for blanketing and purging of production equipment may use nitrogen and the system for blanketing of main cargo tanks may use carbon dioxide. The nitrogen may be produced from compressed air in a membrane-exchange unit. The carbon dioxide may be generated from the combustion of diesel oil in an inert gas generator. Careful handling is required to prevent leakage and spillage of nitrogen because the very low temperature of the escaping nitrogen impinging on a surface of steel components can cause fracture. Also, unlike trading tankers, ship-shaped offshore installations would not share a common inert gas system for the inerting of the slop and cargo tanks.

In the control of overpressure and fast-flow velocities, vent systems are deployed. For cargo handling in tanks, a larger venting system is needed, although a smaller venting system will be used for hydrocarbon production facilities. On a new-build offshore unit, the use of a single venting system that can work both for cargo handling and hydrocarbon production may be considered, but two separate systems are often used for a tanker conversion.

There are three types of drain systems associated with production: closed drains, hazardous open drains, and nonhazardous drains. The closed-drain system gathers the drains from all piped hydrocarbon drain points into a slop tank collection system that can be located below the topsides deck level to achieve the drainage by gravity. In deck drains, the possibility of spillages and leaks must be dealt with. The hazardous open-drain systems may use a trap system to store small spillages that can then flow into a hazardous drains collection tank. In there, the drainage then may be decomposed into oil and water. Recovered oil may be returned to the process, and the oily water may be directed to the slop tank; the separated water may be disposed of overboard providing the right conditions are met. Traps must be cleaned and inspected on a regular basis to avoid any debris accumulation.

The nonhazardous drains system gathers seawater, rainwater, washdown water, and deluge into the slop tank collection system, also through traps. Large volumes of water overflowing the traps can usually be readily disposed of overboard because of its nonhazardous quantities.

Provisions must be made to offer full flexibility for tank inspections and personnel entry into tanks with the highest levels of safety assured. Considerations for heating, cooling, airflow, and other types of systems for the purpose are described in Section 7.2.

7.2 Utility and Support Systems

The utility and support systems available in the topsides facilities may include the following:

- Fuel gas system
- Heating and cooling medium systems
- Starting air/atomizing air for gas turbines
- Seawater treatment and injection system
- Chemical injection and storage systems
- Flare system
- Nitrogen generation system

In ship-shaped offshore installations, many facilities and systems require diesel fuel to operate, and these include main power generation units (turbines or reciprocating engines), gas turbine drivers for large compressors and pumps, emergency generators, fire pumps, and an inert gas generator for cargo tank blanketing. To prevent problems in the fuel gas system, the diesel must be filtered and centrifuged to remove moisture, sulphur, and particulates. The risk of fuel gas leaks in the utilities areas and machinery spaces must be considered in design.

Appropriate design of the heating medium system may help in optimizing the operation of the installation. Waste heat, for example, coming from the exhaust of power generation turbines, may be captured by waste heat recovery units. This can result in meaningful savings because the turbines usually work continuously, thus providing a steady source of waste energy. The cooling medium system is required to cool down the heated fluids; the coolant employed is usually seawater.

7.3 Safeguard Systems

Various systems are required to control and safeguard the operational integrity of the overall installation. These include, for example, the emergency shutdown system (ESD) and the fire and gas control system. The safety systems are controlled as for monitoring, normal operation, startup, and shutdown from a centralized control room usually located in the accommodations unit.

The design of safety systems must comply with various high standards of marine practice and detailed, and often prescriptive, classification and regulatory requirements. Design considerations related to safety systems include the following:

- Provision for smoke and gas detection
- Prevention of smoke ingress into accommodation and control areas
- Smoke clearance, especially in accommodation and control areas and along designated escape routes
- Interactions between main fire and gas system
- Ventilation of the temporary refuge and control areas in emergencies
- Provision of required life support during upset conditions
- Implications of hazardous area classification
- Dispersion of gas in areas where gas escapes may occur
- Fire protection and rating of dampers and ductwork
- Controls, control stations, and control system configuration and complexity

- Position of inlets and exhaust relative to hazardous areas
- Pressurization of spaces and airlocks
- Equipment redundancy and spares

For fire protection, a combination of active and passive protection measures is usually used for ship-shaped offshore installations. Active protection measures are generally provided by fire water (e.g., deluge, sprinklers, monitors, hose reels), foam, CO_2 blanketing, and portable extinguishers. Passive protection measures are provided by fire-rated partitions and fire-resistant coatings applied to primary structural components, bulkheads, decks, frames, equipment foundations, and pressure vessel shells/saddles/skirts. Fire protection should also be provided to flare tower and turbine exhaust supports. The type and extent of the fire protection methods can be affected by the results of fire, heat radiation, and other safety-related studies.

When selecting appropriate materials and features used for passive fire protection, their properties and abilities related to the following need to be considered:

- Withstanding the impact of firewater deluge and fire hose jets
- Remaining intact after an explosion
- Withstanding jet fires
- Being noncombustible
- Accommodating the movement of structural components and equipment affected by the flexing of the hull
- Imperviousness to seawater, firewater, hydrocarbons, and corrosive chemicals
- Chemical inertness to prevent deterioration over time
- Absorbing impacts from tools without spalling

The fire and gas system will contain many detection devices for gas, flame, and smoke. Fire detectors may be supplemented by thermal imaging cameras or closed circuit television. It is important to ensure that the design of the fire and gas system and the monitoring and control hardware must be made by vendors with a proven track record in design, manufacturing, testing, delivery, installation, and commissioning of such systems offshore.

Reference

J.K.Paik & A.K.Thayamballi, Ship-shaped Offshore Installations, pp.318–327, Cambridge University Press, 2007

Home Work Problem

1. Find the meaning of the following words or phrases by consulting the dictionary.

fiscal	inhibitor	biocide	inerting
manifold	isolation	leak	pollution
dehydrate	pertinent	purge	spillage
Impinge	slop	reciprocating	exhaust
Safeguard	ingress	hazardous	implication
redundancy	intact	deterioration	spall

2. Put T or F for the following sentences whether these are true or false, respectively, according to the facts in the passage of this unit.

- For reservoir support during production in the field, high-pressure water usually needs to be injected to the reservoir.
- The water injection will be carried out via high-pressure reciprocating pumps.
- Leakage and spillage of nitrogen impinging on a surface of steel components can cause fracture.
- The hazardous drains system gathers seawater, rainwater, wash-down water, and deluge into the slop tank collection system, also through traps.
- In the control of overpressure and fast-flow velocities, vent systems are deployed.

- The design of safety systems need not comply with various high standards of marine practice and detailed, and often prescriptive, classification and regulatory requirements.

3. Answer the following questions.

- What kinds of treatment are needed to the water to be injected into the reservoir?
- Please list three types of drain systems associated with hydrocarbon production.
- Least at least 5 utility and support systems available in the topsides facilities.
- What is the abbreviation of "the emergency shutdown system" ?
- What are the design considerations related to safety systems? Please list them.
- What are three types of drain systems associated with production? Please list them.

UNIT 8

FPSO (Floating Production Storage and Offloading System)

FPSOs are the most prolific floating production platforms. They were developed in the 1970s to produce from smaller, remote fields where pipelines or fixed structures would be uneconomic. The early FPSOs were restricted to mild environments, which allowed the use of tanker conversions. The turret mooring was introduced in 1986, which, in turn, opened the door for the FPSOs into more severe environments. The first new build and North Sea FPSOs were also introduced in that year (see Ronalds and Lim, 1999, for a review of FPSO history). The North Sea and Brazil are today the primary FPSO markets. In spite of their presence in the world's offshore oil producing regions there was not a single FPSO in the Gulf of Mexico as of 2004. The US government has only recently approved their use in the Gulf of Mexico (Minerals Management Service, 2002), however as of 2004 no oil company has made a proposal to install one.

The FPSOs are generally ship shaped floaters with provisions for storing and offloading of oil simultaneously. They may be designed to weathervane so that they always face the weather, minimizing roll and heave motions. In benign environments such as West Africa and South East Asia, the FPSO may be spread moored to face one direction at all times. Some FPSOs for Brazil have been designed to semi-weathervane by using a spread mooring with slack aft moorings, giving the vessel the options of some limited weathervaning (Kaster et al, 1997; Portello et al, 2003).

Oil production is through either flexible risers or riser towers with flexible jumpers. The motions of the FPSOs generally prohibit the attachment of rigid vertical risers or catenary risers.

The FPSOs have a large area for setting a deck on the top of the hull. However, many FPSO hulls are conversions and the deck structure may not be designed to carry a processing facility. This needs to be checked carefully before committing to a surplus tanker for conversion.

The converted FPSOs often offer the shortest and the cheapest path to initiating production. Their main limitations include a lack of ability to operate dry trees, and technical feasibility of mooring in very deep water in harsh environments. The turret assembly can become very complicated and difficult to integrate with the hull.

8.1 FPSO Hull Design

There are four principal requirements that drive the size of a typical FPSO (Figure 8.1);

1. Provision of oil storage capacity compatible with the production rate and offloading arrangements, i.e. shuttle tanker turnaround time
2. Provision of topsides space for a safe layout of the process plant, accommodation and utilities.
3. Provision of displacement and ballast capacity to reduce the effects of motions on process plant and riser systems.
4. Provision of space for the production turret (bow, stern or internal), and the amount of hull storage capacity lost as a consequence (new-build or conversion).

Figure 8.2 and 8.3 show FPSOs with a bow turret and internal turret, respectively. Kizomba A, figure 8.4, is an example of an FPSO with a spread mooring and no turret.

As water depths increase, the sensitivity of the FPSO mooring and riser systems to wave frequency motions increases rapidly. The FPSO hull form can be optimized to counter this by varying primarily the length, width, depth, draft and mass distribution.

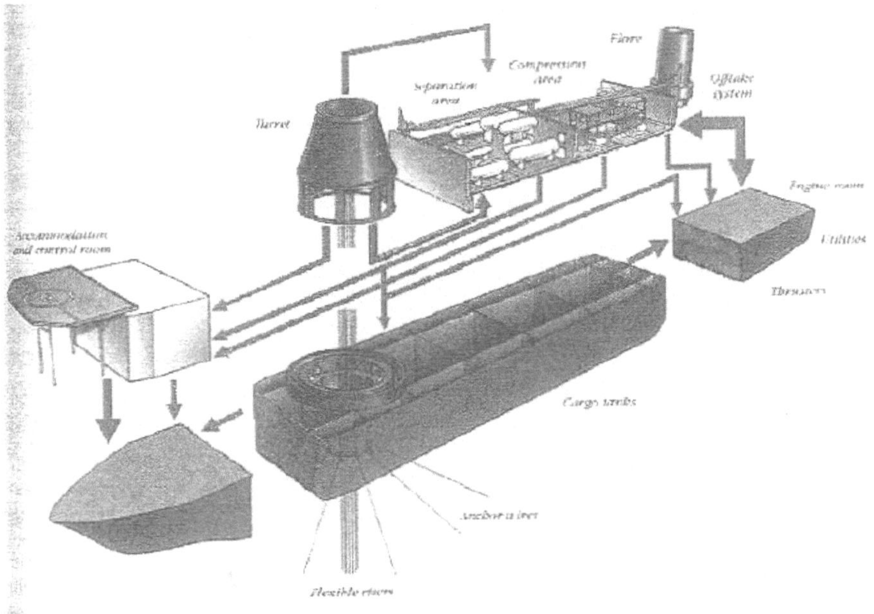

▲ **Figure 8.1** Typical new-build FPSO configuration(Courtesy of J.Ray McDermott, Inc.).

▲ **Figure 8.2** Xikomba with Bow Turret (SBM)

▲ **Figure 8.3** Balder FPSO with Turret Midships (Exxon Mobil)

▲ **Figure 8.4** Kizomba A, Spread Morred FPSO in West Africa (Exxon Mobil)

The hull length can be adjusted with respect to the expected wave lengths to ensure that the extreme environmental conditions do not generate wave lengths located in the range of the peak heave and/or pitching responses, thus reducing the turret vertical motions and associated dynamic loads. The hull form can also be adjusted in order to control rolling and pitching behaviour thus reducing riser bend-stiffener design requirements, particularly for transverse conditions. This optimization also reduces the need for special topsides process equipment design requirements such as baffles for separators.

The hull form can also be optimized to reduce the wave and current actions by varying the length, width, draft and shape of bow and stern. This reduces the vessel loading with respect to incoming waves under transverse conditions, reducing mooring loads and vessel roll. The aforementioned hull form optimization can have a beneficial contribution to the long term operational performance of the FPSO, but this should be carefully assessed against the fabrication cost and Life of Field (LOF) costs.

In the more recent FPSO developments, there have been cases where the cargo carrying capacity of the FPSO is not sized against the initial production flow rate, but one further into the production life where a reduced and more economical storage can be achieved. The initial storage requirements are met by supplementing the FPSO with a Floating Storage unit (FSU) based on a cheap second hand tanker.

The above highlights that a full LOF approach should be taken in deciding both the vessel design and the field configuration. The following sections discuss an example FPSO design for 800,000 BBL storage in deep water.

8.2 Hull Structure

There is a marked difference in the hull arrangement between the new purpose-built FPSO hulls and the converted tanker hulls. The overall configuration for tankers is driven by the need to transport large cargo volumes at a low cost. Tankers have consequently evolved to a length to breadth ratio of about 6:1, which gives a good compromise between the enclosed volume and the resistance to forward motion. The FPSOs are not required to move forward; consequently resistance is not an issue. However, in a weathervaning mode the hull slenderness ratio (length to beam ratio) serves to present a low frontal area to the prevailing environment and assists in the natural weathervaning motion. A low slenderness ratio results in more favorable motions and mooring behaviour over a shorter and more bulky hull. A shorter hull would however offer savings in steel weight and possible cost reductions. Hull breadth to depth ratio comparisons are a different prospect since FPSOs, unlike tankers, are not constrained by maximum draughts, so the typical 2:1 ratio can be optimized. This helps structural design and seakeeping by increasing freeboard whilst permitting greater bow submergence to reduce slamming.

With regard to the issue of the double hull and the double bottom hull design, the former has become the norm for FPSOs, whilst there is little justification at present to impose a double bottom on a statically moored FPSO. It has become common practice to arrange ballast tanks outboard of the central cargo tanks. The ballast capacity depends on the range of operating draughts due to seakeeping requirements and offloading.

Reference

Subrata Chakrabarti, Handbook of offshore Engineering, vol1, pp.448–452, Elsvier, 2003

Home Work Problem

1. Find the meaning of the following words or **phrases** by consulting the dictionary.

in turn	weathervane	benign	slack
aft	catenary	surplus	feasibility
turnaround	accommodation	aforementioned	beneficial
highlight	assess	compromise	seakeeping
submerge	freeboard	slam	

2. Put T or F for the following sentences whether these are true or false, respectively, according to the facts in the passage of this unit.

- FPSOs were developed in the 1990s to produce from smaller, remote fields where pipelines or fixed structures would be uneconomic.
- The motions of the FPSOs generally prohibit the attachment of rigid vertical risers or catenary risers.
- Many FPSO hulls are conversions and the deck structure may not be designed to carry a processing facility.
- As water depths increase, the sensitivity of the FPSO mooring and riser systems to wave frequency motions decreases rapidly.
- There is a marked difference in the hull arrangement between the new purpose-built FPSO hulls and the converted tanker hulls.

- The US government has never approved the use of FPSOs in the Gulf of Mexico
- The FPSOs are not required to move forward.

3. Answer the following questions.

- What are four principal requirements that drive the size of a typical FPSO?
- Please name an example of an FPSO with a spread mooring and no turret.
- What is the full name of the FPSO?
- When was the first FPSO built?
- How could you define the FPSO in general?

UNIT 9

SEMI-SUBMERSIBLES - PART I
HISTORY

9.1 History of the Semi-Submersible

9.1 History of the Semi-Submersible

Semi-submersibles evolved from a drilling vessel type called a "submersible", which operated sitting on bottom in fairly shallow water and provided a working deck well above the highest expected waves (see Lim and Ronalds, 2000). These units transited afloat on pontoons and required "stability columns" to safely submerge to a bottom founded mode of operation. To operate in deeper water, the marine riser was developed and spread moorings were perfected allowing drilling afloat. This first application was with barges, however (e.g. "CUSS I"; see Burleson, 1999 for story of CUSS). To overcome the barges undesirable motions, the basic submersible design of the time was adapted to the floating drilling function. This was the "Bluewater". This was a Shell Oil sponsored development with Bruce Collip as the inventor of record.

Although highly evolved in size and configuration, these semi-submersibles remain fundamentally the same as they originated: a deck ' supported well above the sea by submerged pontoons, with a spread of large columns providing floatation stability. Both its parent, the submersible, and the semi-submersible are officially designated as "column units" (USCG, ABS, etc). The columns are "stability columns" and primarily flotation stability.

Drilling semis may be divided into four generations:

1st. Generation: Before 1971

2nd Generation: 1971-1980

3rd Generation: 1981-1984

4th Generation: 1984-1998

Those most recently built, those since 1998, might be called "5th Generation," but a clear distinction has not emerged.

The first generation consists of a broad variety of configurations developed throughout the 1960s, beginning with the "Bluewater I" and including the notable SEDCO 135 designs and the variously configured ODECO designs (e.g. "Ocean Queen"). With the exception of the SEDCO 135, these designs all featured an array of multiple pontoons (replacing the bottom support mat of the submersible parent). Besides large diameter stability columns at the extremities, they also had many slender interior columns to support the working deck - but no diagonal trussing. The main element of global strength was the pontoon. Conversely, the SEDCO 135, a 3-column design, evolved from a Transworld submersible design and employed independent columns and footing tied together with a trussed space frame. The Forex Neptune "Pentagone 81", a 5-column, independent footing, trussed space frame design, is the culmination of this period. It is a true space frame, without trusses in a bent and also is an early user of hull-type superstructure.

The main elements of this period affecting design was a lack of technology exchange, a lack of consistent design philosophy, and the fact that no single design exhibited a complete understanding of the important principles of semi-submersible design. This period is more of historical interest than anything else. Nevertheless, some of the better 1st generation designs continued to be built well into the 1970s.

It is interesting to note that, as many of the 1st generation semi-submersibles became noncompetitive as mobile drilling units, they found new lives as floating production systems (FPS) of these, the "Transworld 58", was the first true FPS and deployed in the North Sea(1984). More recently (1995), another, the "Glomar Biscay II" (ODECO "Ocean Victory Class"), was converted to an FPS for the Enserch Garden Banks 388 Development.

The 2nd generation produced the majority of the units built. In addition to better technology exchange, this generation was stimulated by competition from drillship. It became evident that a better mobility was required and that eliminating the transverse pontoons was important. Perhaps inspired by the aborted "Mohole Project" Semi-submersible (National Science Foundation, Deep Ocean Drilling

Project of the late 1960s), it became evident that two pontoons would be sufficient. However, this would require a different structural system. The obvious solution was evident in the SEDCO 135 and other similar designs and a truss system was employed which included transverse braces between the columns, and the various arrangements of diagonal bracing. An improvement in the hydrodynamic performance did not go unnoticed. Therefore, the principal feature of the second generation is the twin pontoon configuration and the high mobility it provides.

Although incomplete, due to budget cutting, the Mohole Project was a semi-submersible under construction (see Burleson, 1999). It was nevertheless a watershed of technology. The structural system was somewhat unique and maintained almost entirely by a hull-type superstructure, integrally built into the tops of the stability columns. It did have a shallow truss system with diagonal bracing, but did not have the low, horizontal transverse typical of later designs.

The "Zephyr/Marge Class," drilling units of the early 1970s, emulated the Mohole structural system but, upon entering service in the North Sea, the "Margie" suffered failures in the shallow truss. This event proved the necessity of the horizontal bracing and also the viability of the hull-type superstructure, without which, the "Margie" would have broken up. Another was the Offshore Company's "Chris Chenery." Without the horizontal transverses or the shallow truss, it had a hull-type superstructure, albeit heavily reinforced. It too demonstrated the strength of the hull-type super structure. The "Chris Chenery" and its single sister both have since been fitted with horizontal transverses to extend the fatigue life.

While no single design represented a complete understanding of the design principals, most 2nd generation semi-submersibles were relatively well designed from a performance point of view. An important aspect of this period was a higher level of dissemination of the design and performance knowledge. It is not a coincidence that the first Regulations and the ABS Classing Rules were published in the late 1960s and that the OTC began in 1969. Most notable of this period, if only by its numbers, is the Aker H-3.0 design. Other notables include the Pacesetter

(Friede & Goldman) and the SEDCO 700 classes.

The demarcation between the 2nd and the 3rd generation is rather sharp. In 1979, only two units were delivered and in 1980 there were none. By the time building resumed, the tragedies of the "Alexander Kjelland" and the "Ocean Ranger" had been reflected in new rules, performance requirements had increased, and costs had doubled. While the loss of the "Alexander Kjelland" revealed a structural redundancy flaw of a 1st generation ironically, the "Ocean Ranger," was a 3rd generation design, both configurationally and structurally, or least a precursor.

Beside the increased size, payload, and higher standards of redundancy, the main features of the third generation are the continuation of the twin pontoon form, the use of hull-type superstructure, the well-designed brace connections, and a generally thorough understanding of the design principles of semi-submersibles.

Notable 3rd generation designs are the "Bingo", "Ocean Odyssey" (an improved "Ocean Ranger”), "Scarabeo 5" and the "Zapata Arctic". It should be noted that a number of enhanced 2nd generation designs were built during this period, particularly the Sedco 700 and Pacesetter Classes and even a few 1st generation designs.

The 4th generation is rather difficult to define and is small. The most notable designs are the GVA 4500 and its variants (two FPVs: "Balmoral" and "Petro bras XVIII"), the "Henry Goodrich” and the “"Zane Barnes" (now "Jack Bates"). Typically, the 4th generation semi-submersibles are large, suitable for harsh environment operation, and deep water capable.

From a structural point of view, a marker of the 4th generation semi-submersible designs is that they rely fully on a hull-type superstructure with no bracing other than horizontals between the columns. One of the favorable aspects of this configuration is that, by the elimination of bracing, many inspection problems and the fatigue potential they represent are eliminated. The first use of this structural configuration was in the five Penrod semi-submersible by Reineke in the early

▲ **Figure 9.1** Hulls with ring pontoons; "Asgaard B" FPS (Oyvind Hagen, STAT OIL),
"Snorre B" FPS hull (NORSK HYDRO) and Brutus TLP (SHELL)

1970s. However, due to immature connection detailing practice, some of these suffered from brace cracking. Some were fitted with diagonals and others were used in milder environments without any problem. One of these with diagonals fitted, the "Penrod 72" is another obsolete drilling semi-submersible converted to an FPS. First serving as the Placid GC29 FPS in the Gulf of Mexico, it continues to serve as "Petrobras XXIV." The first new-built-FPS is an unbraced configuration with a hull-type superstructure designed by the Swedish shipbuilder GVA and serves as the "Balmoral" FPS in the North Sea.

The "Asgaard B" FPS designed by Aker, shown in figure 9.1, is the largest semi-submersible built to date (at 81,500 t displacement) and a good example of the unbraced full with ring pontoon. This example also utilizes an integrated truss deck which is installed by a floatover. Several FPS conversions have used cross pontoons to considerable benefit regard. While their towing speed is considerably slower, it is sufficient for the one-time deployment required. For reasons unknown, the two GVA designed FPSs are twin pontoon with transverse braces, but several proposed designs by GVA show the closed pontoon arrangement.

Figure 9.2 shows the 4th generation drilling semi-submersible, MDC "Marine 700." This unit was designed for a large variable deck load and large storage volumes for drilling and completion fluids (Moyer et al, 2001). It is designed for a maximum wave height of 32 m (105 ft) and a wave period of 11 to 15s. Box-type elements and transitions provide a minimum of stress concentration at critical connections and an adequate fatigue life of 20yr, based on North Sea criteria. The "Marine 700" has a transit draught of 9.75 m (32 ft) an operating draught of 23.75 m (77.9 ft) and a survival draught of 21 m (68.9 ft). It has an overall length of 119 m (390 ft) and an overall width of 71.4 m (234 ft).

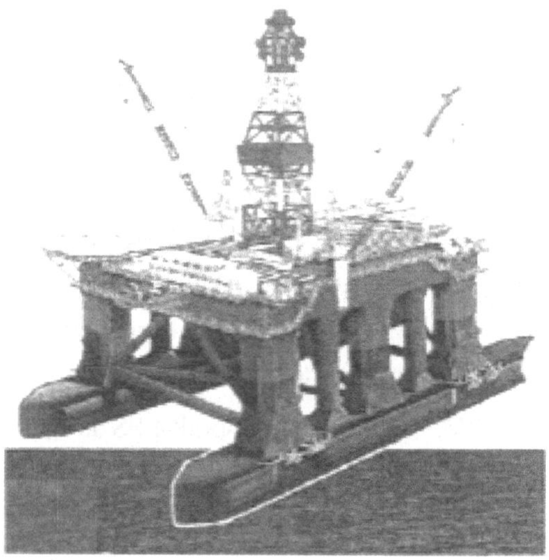

▲ **Figure 9.2** Marine 700 4th generation semi-submersible(Moyer,2001)

▲ Figure 9.3 "Deepwater Nautilus" – 5th generation drilling semi-submersible (PhotocourtesyTransocean)

The “Deepwater Nautilus," owned by Transocean (Figure 9.3) is an example of a 5th generation semi-submersible designed for ultra-deepwater drilling.

Other semi-submersibles have been used for accommodations, derrick vessels, pipelaying, and work vessels. While some are conversions, most are special purpose newbuilds, somewhat unique in their size or arrangements and generally include 3rd and 4th generation designs.

Reference

Subrata Chakrabarti, Handbook of offshore Engineering, vol1, pp.464–471, Elsvier, 2003

Home Work Problem

1. Find the meaning of the following words or phrases by consulting the dictionary.

afloat	extremity	culmination	noncompetitive
abort	watershed	emulate	viability
albeit	dissemination	demarcation	redundancy
precursor	immature	obsolete	draught

2. Put T or F for the following sentences whether these are true or false, respectively, according to the facts in the passage of this unit.

- Both its parent, the submersible, and the semi-submersible are officially designated as "column units" (USCG, ABS, etc).
- The drilling semis built since 1998 might be called "4th Generation" .
- The SEDCO 135, a 3-column design, evolved from a Transworld submersible design and employed independent columns and footing tied together with a trussed space frame.
- The "Transworld 58, was the first true FPS and deployed in the North Sea(1984).
- The "Chris Chenery" and its single sister both have been fitted with vertical transverses to extend the fatigue life.
- The 4th generation semi-submersibles are large, suitable for harsh environment operation, and deep water capable.

- The "Deepwater Nautilus," owned by Transocean is an example of a 4th generation semi-submersible designed for ultra-deepwater drilling.

3. Answer the following questions.

- How can the drilling semis be divided into four generations? List them.
- What are notable 3rd generation designs of semis? Please name them.
- What are notable 4th generation designs of semis? Please name them.
- What is the first new-built-FPS which was used in the North Sea?

UNIT 10

SEMI-SUBMERSIBLES - PART II DISTINCTIONS AND DESIGN

10.1 Distinctions between a MODU Semi-submersible and an FPS

10.2 Semi-submersible Design

10.1 Distinctions between a MODU Semi-submersible and an FPS

Most semi-submersibles begin life as a "Mobile Offshore Drilling Unit" (MODU). The function of a MODU and the operating patterns of a MODU are decidedly different from those of an FPS. Fortunately, the configuration and much of the equipment and systems are useful. A MODU has only the mission functions, drilling and well service, but is amply provided with support functions.

A particularly costly support function of a MODU is its mobility that is not required of an FPS. Mobility is the primary reason for the twin pontoon configuration. However, the twin full configuration makes global strength more difficult to provide.

Mobility in general, and quick deployment in particular, is a major factor in the mooring design. The MODU moorings are rarely suitable for long term applications. For deep water FPS applications moorings are a particularly significant system cost component, making minimization of wind and current load a more important design priority than is for a MODU.

Another important MODU distinction is its maintenance. Mobility provides considerably more favorable circumstances for maintenance. These and more subtle differences can be reasonably dealt with in a conversion and likewise need to be recognized in the design for new construction. For strength reasons discussed later, the production semi-submersibles tend to the closed array pontoon configuration.

10.2 Semi-submersible Design

Semi-submersibles consist of a deck, multiple columns and pontoons. They are "column stabilized", meaning that the center of gravity is above the center of buoyancy, and the stability is determined by the restoring moment of the columns.

This contrasts with the spar platform, which achieves stability by placing the centre of gravity below the centre of buoyancy, and the TLP, whose stability is derived from the tendons.

The design of semi-submersibles depends on these principle considerations which are somewhat generic to floater concepts:

- Weights and CG's (cycle of steadily improving estimates)
- Hydrostatics; tank capacities
- Intact and Damaged Stability
- Wind Forces (stability and mooring loads)
- Current forces (mooring loads)
- Ballast System Performance ,
- Motions (seakeeping; drift and low frequency mooring loads)
- Global Strength
- Fatigue

Before initiating the design, there should be a definitive Functions List (e.g. production, drilling, quarters), a Systems Summary, and an Equipment List (mission and support). Trial equipment and systems layouts should be made and coordinated with any constraints needed in the initial design. The constraints might include, for example:

- Maximum lightship draft for quayside outfitting,
- Maximum beam for canal transit or dry transportation
- Maximum lightship weight and VCO envelope for dry transport,
- Environmental criteria for operations, transit and survival,
- Maximum lateral eccentricity of the deck load which needs to be trimmed,
- Maximum allowable motions (angles, accelerations) for each given environmental and load condition,
- Applicable rules and standards.

There may be different constraints for various load cases: operations, transit, survival, and installation. These should be identified in order to be able to check the configurations for each case.

Weight estimates need to be made of all permanent payload and variable loads, including equipment and systems outfit for the functions (drilling, processing, utilities, quarters, flare. etc.). The equipment weights are to be supplanted by vendor equipment as it becomes available. The proforma arrangements and calculations should be developed to support the outfit estimates (piping, access, corrosion protection, etc.). In addition, variable load requirements in amount, distribution, and with respect to the operating state should be firmly established. And, if it matters, the installation weight-states for permanently sited platforms should be determined.

Variable weights should be identified for each load case as illustrated in the following table.

	Operating	Transit	Survival
Variable Ballast	x	minimal	x
Drilling Fluids	x		x
Process Fluids	x		
Drill Pipe	x		x
Casing	x		x

The recommended overall process is to execute initial design to conclusion, later addressing the specific shape (pontoon cross-section and ends, column section, flare, etc.) and the use and arrangement of bracing. More than one alternative might be kept for further consideration. Immediately following this, rigorous hydrostatic and intact stability analyses, revised weight estimates, and the motion analyses should be made for each retained alternative. From these results, a choice should be made and a final, definitive principal dimensions sketch should be produced. The outcome of this, particularly the column size and arrangement, should be more a matter of weight, buoyancy, and intact stability. The damaged stability is addressed later by subdivision. Adequate motions response should already have been achieved through refined cross sectional shape of the pontoons. Choices regarding station keeping loading would be resolved with a limited number of alternative initial design variants with addressed and rigorous analyses of the simple model forms. At this point, work should begin on the basic scantlings with the specific objective of establishing the framing system, initial scantlings, and refinement of the steel weight estimate. Concurrently, damaged stability analysis should also proceed in order to determine internal subdivision, internal weight distribution and piping requirements. Global strength analysis can only begin after the basic structural design (framing and scantlings) is largely complete. Fatigue analysis is reserved for detailed design.

Reference

Subrata Chakrabarti, Handbook of offshore Engineering, vol1, pp.464–471, Elsvier, 2003

Home Work Problem

1. Find the meaning of the following words or phrases by consulting the dictionary.

amply	mobility	deployment	priority
pontoon	stabilize	tendon	generic
hydrostatics	fatigue	definitive	constraint
draft	lateral	eccentricity	trim
supplant	outfit	rigorous	concurrently
intact	subdivision	variant	scantling

2. Put T or F for the following sentences whether these are true or false, respectively, according to the facts in the passage of this unit.

- The MODU moorings are suitable for long term applications.
- The damaged stability is addressed later by subdivision.
- Weight estimates need to be made of permanent payload only.
- Mobility provides considerably more favorable circumstances for maintenance in MODU.
- The function of a MODU and the operating patterns of a MODU are not much different from those of an FPS.

3. Answer the following questions.

- What does the abbreviation 'MODU' mean?
- What is the major difference between MODU and FPS?
- What is the meaning of the words "column stabilized" ?

UNIT 11

TENSION LEG PLATFORMS - PART I INTRODUCTION

11.1 INTRODUCTION

11.1 Introduction

Tension Leg Platforms have been used exclusively as production and drilling platforms, with the exception of the "East Spar" platform, which is a control buoy.

Like a semi-submersible, the TLPs consist of columns and pontoons. The unique feature is the mooring system, which consists of vertical tendons (sometimes called "tethers"), which restrain the heave motion. Figure 11.1 shows the configuration and terminology applied to the TLPs.

The previous section also applies to tension leg platforms (TLPs). The reader is therefore advised to have read Section 7.7 prior to continuing this discussion. The TLP is nevertheless fundamentally different in many respects, and this section will concentrate on these differences and note the contrast and the distinctions that make the difference.

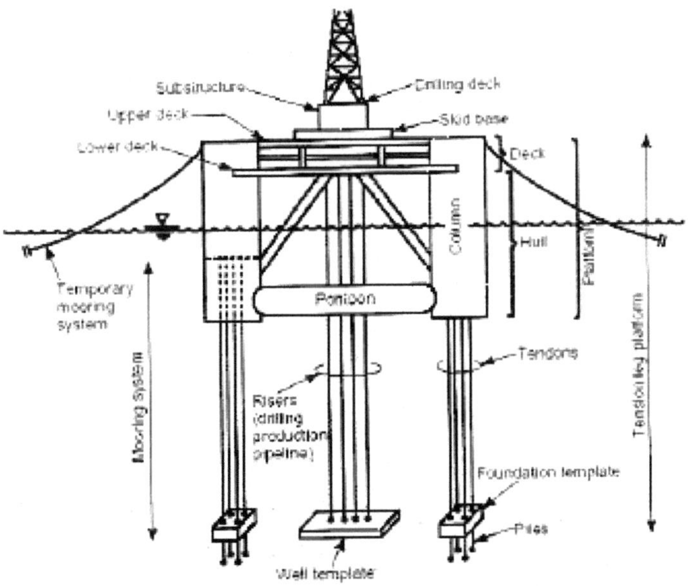

▲ Figure 11.1 Tension leg platform terminology (APIRP2T. Reproduced with permission of the American Petroleum Institute)

The emphases in this section is on the design of the TLP as a platform particularly with regard to its sizing, proportions, tendon arrangement, and the integration of many diverse, important design aspects. Moreover, this section addresses the issue of making the key fundamental decisions of overall design. Analysis is left to confirming those decisions and to refining the choices. It is hoped that, instead of precise answers for a wrong design, good answers for a good design prevails.

As with the semi-submersible, the main problem in TLP design is to address the specific functions and rational sizing. Even more so than semi-submersibles, the construction program consideration is a salient design issue. Here too the approach is to initiate the design with a straight forward process for sizing, and to discourage multiple trial-and-error analyses with inappropriately rigorous and detail methods.

The initial design(s) should be represented in a sketch indicating the chosen configuration and all important dimensions to be used in the preliminary analysis and design activity, including tendon arrangements. The internal hull subdivision need not be included, but a reasonably definitive conceptual of the deck should. The initial design should include the best working weight estimate, hull displacement, and hydrostatics available.

The key analytic areas for preliminary analyses of the initial design for a TLP include the following:

- Weights and CG's
- Wind Forces
- Current forces
- Global Performance Analyses
  - Motions
  - Drift force
  - Tendon tensions
- Global Strength

With a well-executed initial design, a model can be quickly established for the above analyses to rigorously proceed in parallel without the need for additional major design iterations. Results for the preliminary analyses, based upon the initial design, can be used for a reasonably conclusive revision, this being the preliminary design. The initial design is also adequate for the beginning of specialized subsystem (topsides, tendons, riser, installation, etc.) design and analyses to proceed concurrently. The following is about how to develop such a model.

Unlike semi-submersibles, hydrostatics and stability are not salient design issues for a TLP, although they are important considerations for addressing transport and installation. Eventually, in the initial design process, internal subdivision does need to be addressed. For a TLP, internal subdivision is primarily based upon avoiding excessive reduction (or increase) of tendon tension from internal flooding. However, an additional subdivision may be required for transit/installation damaged stability.

The design spiral for the TLP, represented in figure 11.3, indicates that the parameters are continuously updated for more refined analysis or as criteria are changed. For a TLP, the preliminary design, as outlined above, should require much less refinement of the initial design than with a semi-submersible. The refinement should mainly be confined to minor changes in column diameter, pontoon dimensions, and initial tendon tension, and should probably not require a change in draft. With timely execution of the deck layout work and weight estimates, it is possible to conclusively fix the column spacing as well as vertical dimensions with the initial design.

Unlike the semi-submersible, the TLPs to date have been exclusively used for permanently sited production systems, most with drilling or workover functions. They have fewer functions to consider and therefore limited configurational variants.

Although it is implicit in any design, the construction, transport and installation scenario is a particularly important aspect of TLP design with considerable impact on certain design choices. Usually, the hull and deck are separately fabricated, with either an inshore hull-deck mating or an offshore heavy lift. In either case transit ballast to the operating draft, stability, and installation of the tendons are key issues in the design of the TLP. Many of the comments made for semi-submersible design are useful considerations addressing TLP transportation and installation issues. In particular, weight, vertical center of gravity, and stability are design constraints. Otherwise, once installed stability is not a design factor, although weight and weight distribution is very much so.

As with the semi-submersible a definitive Functions List, Systems Summary, and Equipment List (all: mission and support) should be developed concurrently to the initial design work. This should likewise include a trial equipment and systems layout, with particular emphases on how and when the deck and hull are joined, identifying any design constraints needed in the initial design. Deck layout must address the number of levels, minimum spacing between columns, future addition/removal of equipment, and its installation. In addition, a weight estimate for all equipment, systems outfit, and the deck structure should be made, particularly as a necessary element of installation planning. Pending vendor equipment weight data, weights should be estimated from the historical data. The weight estimate should be based upon proforma arrangements and calculations and include outfitting (piping, access, corrosion protection, etc.). For the design operating states, variable load and its distribution should be firmly established. The conclusion of the initial design should reflect the preliminary topsides work, particularly with regard to column spacing, height of the deck, its weight, and the wind profile.

Apart from hull steel-weight estimates, development of the hull scantlings and framing system decisions should be deferred until the preliminary design is completed. However, a preliminary global strength analysis should be completed as a guide to the development of scantlings.

Before completion of the preliminary analyses and design, a re-evaluation of the installation process should be made, particularly with regard to its stability and ballasting. Any adjustments found necessary to facilitate construction and installation should become evident at this point and be incorporated.

Reference

Subrata Chakrabarti, Handbook of offshore Engineering, vol1, pp.501–507, Elsvier, 2003

Home Work Problem

1. Find the meaning of the following words or phrases by consulting the dictionary.

as of	pontoon	tendon	tether
with regard to	confirm	prevail	as with
salient	rigorous	iteration	spiral
refinement	conclusive	workover	implicit
scenario	ballast	draft	concurrently
constraint	vendor	proforma	outfit
apart from	scantling		

2. Put T or F for the following sentences whether these are true or false, respectively, according to the facts in the passage of this unit.

- Like a semi-submersible, the TLPs consist of columns and pontoons.
- Hydrostatics and stability are salient design issues for a TLP.
- For a TLP, the preliminary design, as outlined above, should require much more refinement of the initial design than with a semi-submersible.
- The TLPs to date have been exclusively used for permanently sited production systems, most with drilling or workover functions.
- Once installed stability is not a design factor for TLP, although weight and weight distribution is very much so.

3. Answer the following questions.

- Please list all the key analytic areas for preliminary analyses of the initial design for a TLP.

UNIT 12

TENSION LEG PLATFORMS – PART II FUNCTIONS AND CONFIGURATIONS OF TLPs

12.1 Functions

Unlike the semi-submersible, the tension leg platform is a more recent development and is exclusively used as a permanently sited production platform. Mobility is not an issue. With a minimal range of functions, TLPs have very little configurational dependence upon function. Size does present certain operational logistics problems and permanent siting reduces the maintenance options. Also, TLPs inherently do not lend themselves to petroleum storage as an option.

Besides the mission and support functions, the essential function of a TLP is to support a payload above the highest waves. More specifically, the hull is to provide buoyancy, both for the support of weight and to provide tendon tension. It should also be tall enough to give the deck wave clearance in all modes of operation. Tendon tension has as much influence on hull size as the payload. Functional requirements of the deck and well systems can also influence column spacing.

Unlike a semi-submersible, it is the columns that are the principal source of buoyancy, with the pontoons functioning more as structure. The pontoons do have a size relationship with the columns in regard to hydrodynamic force. Also, unlike the semi-submersible, in the operational state, the TLP employs very little ballast. Ballast is provided to even loading between tendons and also to offset unused payload capacity. Beyond this like the semi-submersible, the size, submergence, proportion, and spacing of the columns and pontoons are major factors in the hydrodynamic performance.

While the heave motion of a semi-submersible is a salient design issue, vertical motion of a TLP is far less of an issue and entirely different. While the TLP does not heave, it will undergo set-down with offset. Like a semi-submersible, the TLP is laterally compliant and will surge, sway and yaw. In both platform types, there is relatively little design-wise that be done to affect lateral motions, although steady offset can be minimized by increased tendon tension.

12.2 Configurations

The three main configurational components are:

- pontoons
- stability columns
- deck

Figure 12.1 illustrates four TLP configurations. The Brutus Platform is a typical example of a four column, "classic" TLP. Unlike semi-submersibles, TLPs do not employ space frame bracing. One reason is, given the closed pontoon array, there is less need of them.

Only recently have TLP configurations begun to significantly evolve. Except for the Hutton platform, in the late 70s, which was a rectangular, 6-column unit, all TLPs until the late 1990s were square 4-column units. Three-column designs

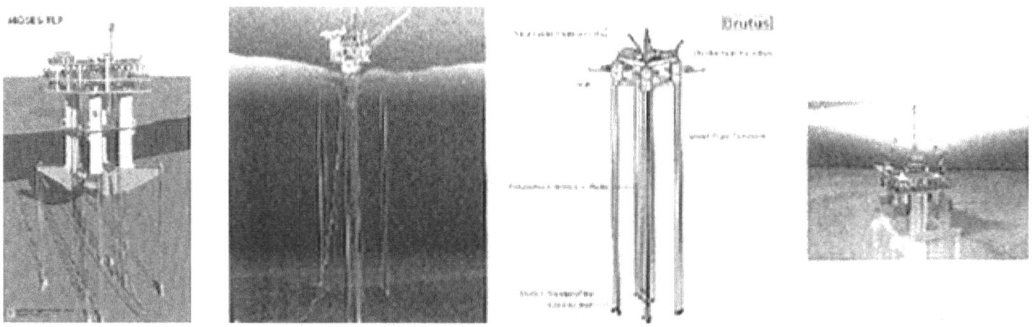

(a) (b) (c) (d)

▲ **Figure 12.1** Example TLP configurations;
A - Moses TLP (MODEC)   B – Seastar TLP (SBM - Atlantia)
C – Brutus TLP (Shell)   D – ETLP (ABB - Luminus)

have been proposed, but not built. Since the late 1990s, four single column designs (Seastar figure 12.1B) and two close-clustered, multi-column designs (Moses figure 12.1A) have been built. Both designs have external, radial pontoon arrangements. Another innovation is removable sponsons (external column additions) for installation. Yet another factor is the short, external, radial pontoons at the columns, connecting the tendons (figure 12.1D). These extend the tendon spread and reduce maximum tendon tension, as well as reduction in deck spans.

Other configurational choices involve deck design and connection of the deck with the columns. The pontoon-to-column connection configuration is another, although this is driven primarily by structural design considerations. This is largely the case whether to adopt the circular or the square column and/or pontoon cross-sections and, in the case of square, whether to have radiused corners.

The mission functions and associated support functions generally do not significantly influence configuration. The well-system arrangement can have impact on column spacing and deck design. However, construction and installation are more likely to influence configuration. This is evident in some recent construction. Mating of the deck will have an importance on the deck configuration, particularly its interface with the columns. Many of the innovations have addressed installation issues, tendons, and the deck particularly risers and well systems.

Decks of TLPs (and some production semi-submersibles) are unique. Virtually all TLP decks are separately built from the hull (often on a different side of the world) and joined later, either at dockside, offshore, or in a separate, sheltered location. An underlying issue with this uniqueness is more administrative than technical. Except at the very early design stage deck and hull are separately designed and managed. Because of their functional and complexity, and an extensive design history, design of TLP decks is very different from those of other floating systems. This is also true of the FPSO production decks.

To date, classic TLP decks have always been long span, four-point corner support, open trusses. As such they are multilevel structures. The single column and close-clustered, multi-column designs have featured deck designs very much like those of fixed platforms, with a four-point support as if a 4-pile jacket supports it. A very similar arrangement has been used on the spars. The installation process is with an offshore heavy lift very much as it is with fixed platforms. Preference of the open truss is both to avoid ventilation requirements (as in a hull-type deck) and to facilitate existing fabrication and outfitting methodology. Outfitting makes extensive use of distinct, separately built modules. Besides the intrinsic structural function, to transfer the weight of the deck and its loading to the columns, the deck is also a part of the overall global strength system and must be considered so. Decks also have significant installation considerations to address.

With TLP hull installation benefitting from minimization of topside weight, usually some parts of the deck system are placed after installation of the TLP. The deck-to-column interface is of considerable importance, with many options for load transfer and securing, and many being proprietary. These include the welded and mechanical connections, some shimmed to achieve a favorable pre-load. This connection could be an integral, full moment connection, as in most semis, but this has not been preferred.

As noted above, with the exceptions noted, the 4-column configuration has dominated the design choices until the late 90s. These all have the closed array pontoon arrangement. The reader is referred to the corresponding discussion for semi-submersibles for addition attributes of this arrangement. It should be noted, however, that the column to pontoon side relationship is very different in TLPs from semi-submersibles, not to mention important differences in load patterns.

For reasons discussed regarding decks, there has been an avoidance of any functionality (equipment, tanks, etc.) within the columns. There has been a small tendency away from this especially the storage of liquids and the handling of saltwater. Otherwise, the columns and pontoons consist of vast empty spaces with much watertight subdivision.

Reference

Subrata Chakrabarti, Handbook of offshore Engineering, vol1, pp.501–507, Elsvier, 2003

Home Work Problem

1. Find the meaning of the following words or phrases by consulting the dictionary.

mobility	issue	attribute	not to mention
lateral	innovation	sponson	methodology
intrinsic	proprietary	shim	be referred to

2. Put T or F for the following sentences whether these are true or false, respectively, according to the facts in the passage of this unit.

- Virtually all TLP decks are separately built from the hull and joined later.
- Design of TLP decks is very different from those of other floating systems.
- The mission functions and associated support functions generally significantly influence configuration.
- Mobility is a big issue for TLP construction.
- TLPs inherently lend themselves to petroleum storage as an option.
- In the operational state, the TLP employs very large ballast.
- Functional requirements of the deck and well systems cannot influence column spacing.

3. Answer the following questions.

- What is the reason that TLPs do not employ space frame bracing unlike semi–submersibles?
- What is the essential function of a TLP?
- What are the three main configurational components of a TLP?

UNIT 13

SPAR

13.1 Introduction

13.2 History of Spars

13.3 Spar Description

13.4 Spar Riser Systems

13.5 Spar Mooring

13.1 Introduction

The Spar concept is a large deep draft, cylindrical floating Caisson designed to support drilling and production operations. Its buoyancy is used to support facilities above the water surface [Glanville, et al 1991; Halkyard 1996]. It is, generally, anchored to the seafloor with multiple taut mooring lines.

In the mid-seventies, Shell installed an oil storage and offloading Spar at Brent Field, in the NorthSea. The hull is 95ft (29m) in diameter, necks down to 55ft (17m) at the water plane, and the operating draft is 357ft (109m). Agip installed a flare Spar off WestAfricain1992. TheSparis233ft (71m) long, with a diameter of 7.5ft (2.3m), which tapers to 5.5ft(1.7m) through the water plane. In 1993 Shell installed a loading Spar at Draugen. The hull diameter is 28ft (8.5m) and the operating draft about 250ft (76m).

▲ **Figure 13.1** Oryx/CNG Neptune Spar(Kerr McGee)

The world's first production Spar was the Neptune Spar installed in 1996 by Oryx Energy Company (now Kerr-McGee) and CNG (figure 13.1). The Neptune Spar has a hull 705ft(215m) long with a 32 x $32ft^2$($10x\ 10m^2$)centerwell and a diameter of 72ft (22m). The mooring system consists of six lines consisting of wire rope and chain (figure 13.1).

As of this writing there are 13 Spars in production or under construction. Figure 13.2 shows the progression of Spars built by Technip Offshore, Inc. Three additional Spars have been built by J. Ray McDermott.

The first three production Spars consisted of a long cylindrical outer shell with "hard tank" near the top to provide buoyancy. The middle section was void, free flooding and the lower section consisted of "soft tanks" which were only used to allow horizontal flotation of the Spar during installation, and for holding fixed ballast, if necessary. Subsequent Spars replaced the middle section with a truss structure to reduce weight and cost, and to reduce current drag. Horizontal plates were included between the truss bays to trap mass in the vertical direction to minimize heave motions. Figure 13.3 shows these two types of Spars, the "classic" and the "truss" Spars.

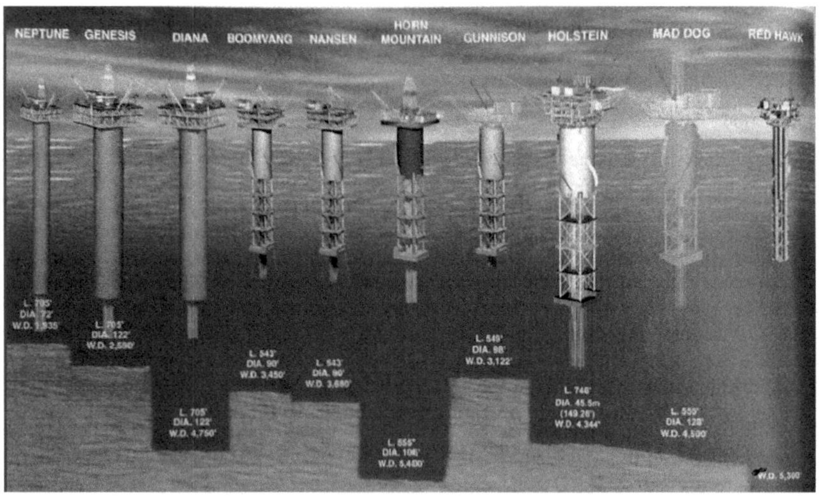

▲ **Figure 13.2** Progression of Spars (Technip Offshore)

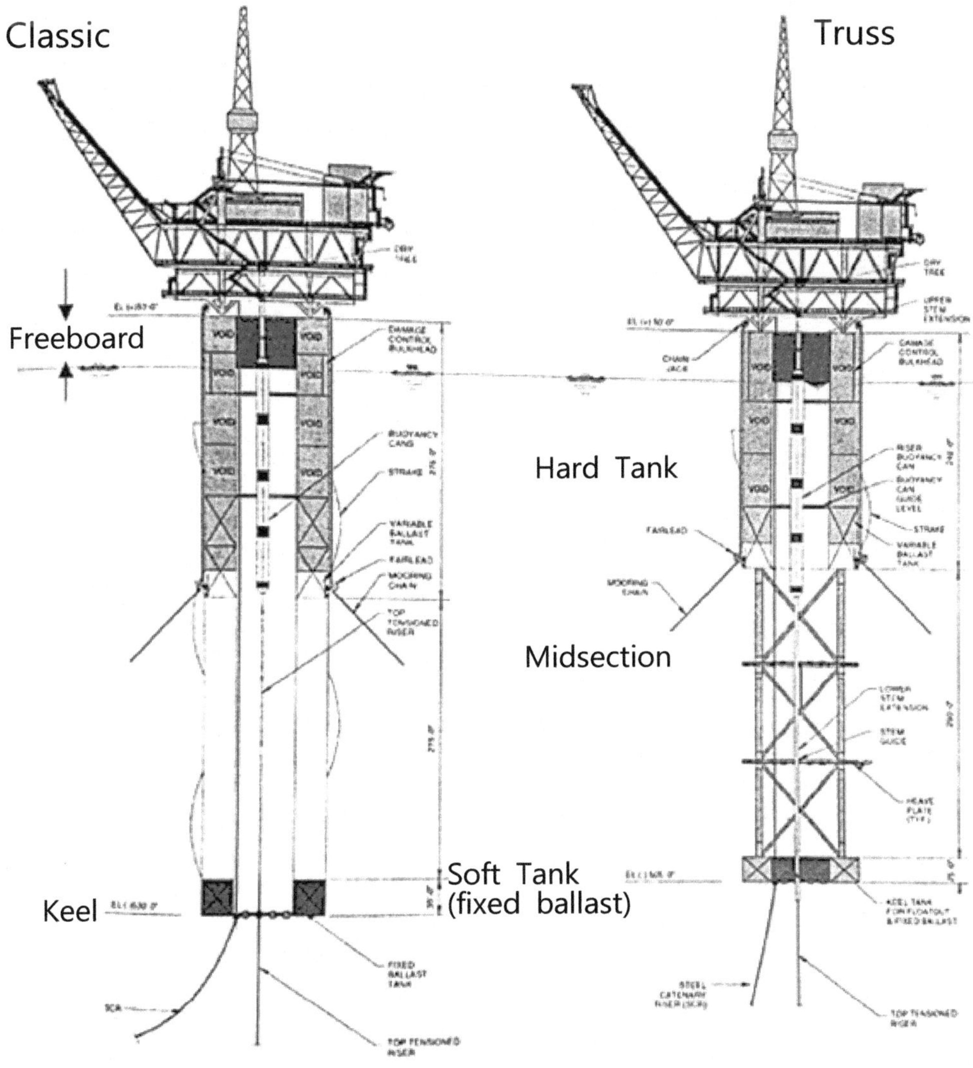

▲ **Figure 13.3** Classic and truss Spar (Technip Offshore)

A third generation "cell" Spar was introduced in 2004. It performs similar to the other Spars, but it is constructed differently. The hull consists of multiple ring-stiffened tubes, or "cells", which are connected by horizontal and vertical plates. This method of construction is cheaper than the traditional plate and frame methods.

Because of the length of a Spar, the Spar hull cannot be towed upright. Therefore, it is towed offshore on its side, ballasted to a vertical attitude and then anchored in place. The topside is not taken with the hull and is mated offshore once the Spar is in place at its site. The mooring cables are connected with pre-deployed moorings.

13.2 History of Spars

Spars have been used for decades as marker buoys and for gathering oceanographic data. The first significant Spar for our purposes is Flip, a structure owned by the U.S. Navy and operated by the Scripps Institution of Oceanography in California(Fisher,1963). Flip was put into service in 1965 and was used primarily for ocean acoustic measurements. Operating draft is about 270 feet . Its diameter is 12ft at the waterline tapering to 20ft over most of the hull. Its heave natural period is 29 s. Although Flip occasionally has been tethered to the seafloor, it is more commonly allowed to drift in the ocean currents.

In the early 1960s, Nippon Telegraph installed a Spar off the coast of Japan to carry a microwave relay station. This Spar was 445ft long with a stepped hull ranging in diameter from 10 to 20ft. The topside structure is a cylinder, 50ft in diameter by 33ft high, with equipment, accommodations and a heliport on top. A four-point catenary mooring of 3 in.-chain connected to a 175 ton clump weights keeps the Spar in place. Operating draft is 330ft. In the mid-seventies Shell installed an oil storage and offloading spar at Brent Field, in the North Sea. The hull is 95ft in diameter ,decks down to 55ft at the water plane, and the operating draft is 357ft. This Spar was designed to store 300,000 barrels of produced crude and to transfer it to bow loading tankers. The oil storage system used a water displacement principle which allowed the tanks to be designed for ambient pressures. The mooring system consists of six lines, each made up of a 1000 ton concrete gravity anchor, 2600ft of 3.5 in. wire rope, and 935ft of 4 in. chain. The top part of the superstructure rotates to allow the tanker to weathervane about the spar. Accommodations, power plant, other equipment and pumps are in the superstructure and a heliport is fitted.

Agip installed a flare Spar off West Africa in 1992 designed to flare 100 million cubic feet of gas per day. The Spar is 233ft long, with a diameter of 7.5ft which tapers to 5.5ft through the water plane. With a draft of 170ft, it is held in place with a four leg chain catenary mooring. In 1993, Shell installed a loading Spar at Draugen. The Spar topsides include rigid boom to which the flexible loading hose is attached. The hull diameter is 28ft and the operating draft about 250ft.

The world's first production spar was the Neptune Spar installed in 1996 by Oryx energy Company (now Kerr McGee) and CNG (Vardeman et al, 1997). Design production rates were 25 mbod and 30 mmcfd respectively. The maximum topsides weight with the workover rig is 5500 tons. Wells would be predrilled with a semi-submersible and completed with a platform workover rig placed temporarily on the Spar. The Neptune spar has a hul 1705ft long with a 32ft by 32ft centerwell and a diameter of 72ft. The six point-mooring system consisted of driven pile anchors, 4.75 in. spiral strand wire rope and chain for the section leading up to the fairleads and onto the hull(figure13.1).

As of this writing there are 14 spars in production or under construction. Figure 13.2 shows the progression of spars built by Technip Offshore, Inc. Three additional spars have been built by J. Ray McDermott.

The Spar, along with the Tension Leg Platform, is the only "floating" platform which up to 2003 has been used for dry trees. The reason for this is that these are the only platforms with small enough heave and pitch motions to allow the risers to be safely and economically supported by the floater.

13.3 Spar Description

Three types of production spars have been built to date: the "Classic" and "Truss" spars shown in figure 13.3, and recently the third generation cell spar (Finn and Maher, 2003). The basic parts of the classic and truss spar include (refer to figure 13.3):

1. Deck
2. Hard Tank
3. Midsection (steel shell or truss structure)
4. Soft Tank

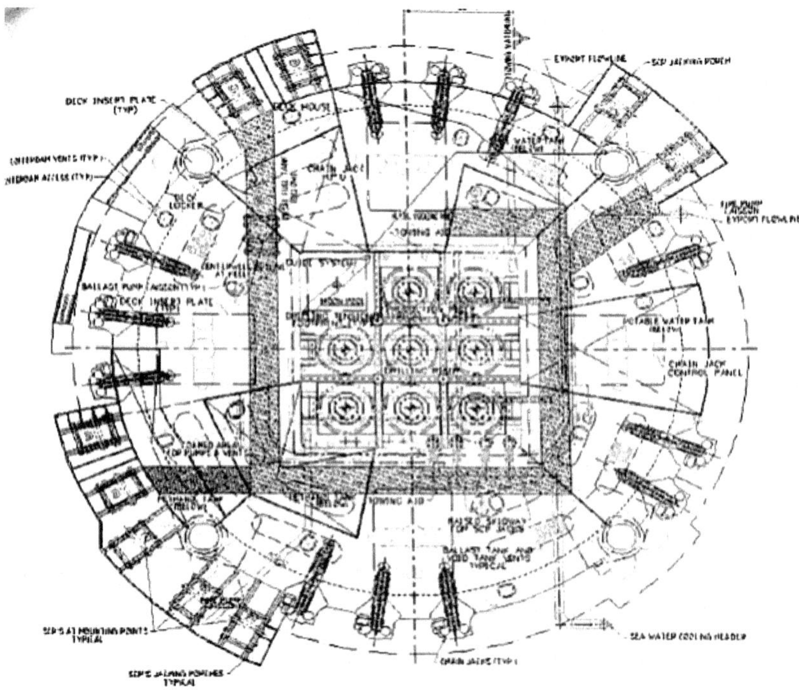

▲ **Figure 13.4** Spar deck arrangement (Technip Offshore)

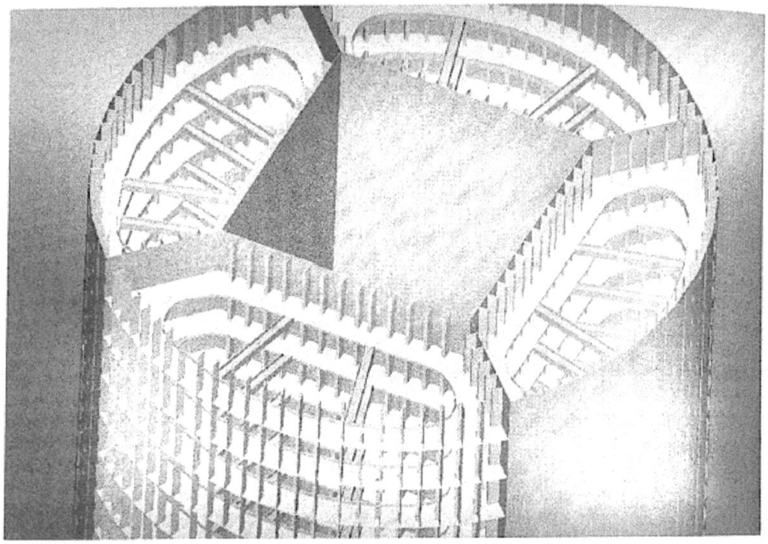

▲ **Figure 13.5** Hard tank structural arrangement (Technip Offshore)

The Topsides **Deck** is typically a multi-level structure in order to minimize the cantilever requirement. For decks up to about 18,000 tons, the deck weight is supported on four columns which join the hard tank at the intersection of a radial bulkhead with the outer shell. Additional columns are added for heavier decks. Figure 13.4 shows the arrangement of the top of a spar with the deck supports. Decks up to about 10,000 tons may be installed offshore with a single lift. Larger decks require multiple lifts.

The **Hard Tank** provides the buoyancy to support the deck, hull, ballast and vertical tensions (except the risers). The term "Hard Tank" means that its compartments are designed to withstand the full hydrostatic pressure (see section of Hull Structure). Figure 13.5 illustrates a typical structural arrangement. The profile is shown in Figure 13.3. There are typically five to six tank levels between the spar deck and the bottom of the hard tank, each level separated by a water tight deck. Each level is further divided into four compartments by radial bulkheads emanating from the corner of the centerwell. The tank level at the waterline includes additional cofferdam tanks to reduce the flooded volume in the event of a penetration of the outer hull from a ship collision. Thus there are up to 28 separate compartments in the hard tank. Typically, only the bottom level is used for variable ballast, the other levels being void spaces.

The **Midsection** extends below the hard tank to give the spar its deep draft. The selection of the spar draft is discussed below. In the early "classic" spars the midsection was simply an extension of the outer shell of the hard tanks. There was no internal structure, except as required to provide support for the span of risers in the midsection. The scantlings for the midsection were determined by construction loads and bending moments during upending. Later spars replaced the midsection with a space frame truss structure. This "truss spar" arrangement resulted in a lower weight, less expensive hull structure. Also, the truss has less drag and reduces overall mooring loads in high current environments.

The **Soft Tank** at the bottom of the spar is designed to provide floatation during the installation stages when the spar is floating horizontally. It also provides compartments for the placement of fixed ballast once the spar is upended. The soft tank has a centerwell and a keel guide which centralizes the risers at that point.

13.4 Spar Riser Systems

Top tensioned risers are placed in a centerwell which runs the length of the hard tank. The risers are typically supported on buoyancy cans which provide the top riser tension instead of hydraulic tensioners used on TLPs. The buoyancy cans are located in guides, which provide lateral constraint. As the spar heaves the buoyancy cans remain in place while the spar moves relative to them. As the spar offsets, the geometry of the risers causes draw down and again the guides allow the risers to slide relative to the hull. The buoyancy can guide system allows stroke of the risers to be relatively large compared to that for typical hydraulic tensioners.

A schematic of the riser arrangement in the centerwell is shown in figure 13.6. The buoyancy cans have an outer shell and an inner stem. The stem has an internal diameter sufficient to allow passage of the riser tie back connector. When in place the riser itself is centralized within the stem. The stem extends above the buoyancy cans to the tree elevation at the production deck level. The risers and the production tree are supported by the stem there. Production fluids from the risers are carried to a production manifold by flexible jumpers, which accommodate the relative movement between the risers and the hull.

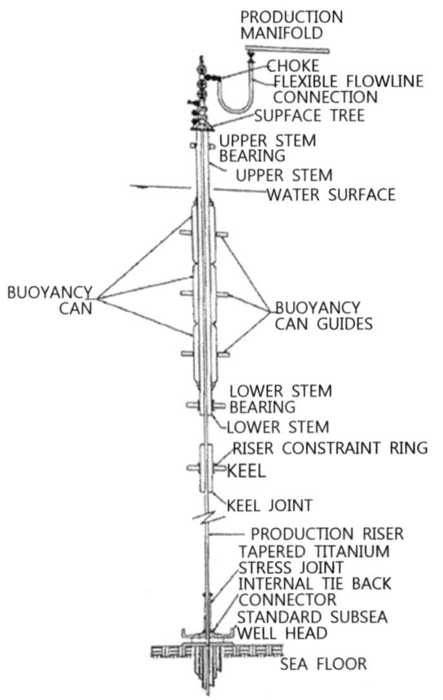

▲ **Figure 13.6** Schematic of riser and buoyancy cans (Glanville et al, 1997)

▲ **Figure 13.7** Buoyancy can being installed (Technip Offshore)

▲ **Figure 13.8** Surface trees on a spar (courtesy Kerr McGee)

Figure 13.7 shows one of the buoyancy cans being installed. Figure 13.8 shows the upper stem, X-mas trees and flexible jumpers.

Drilling risers on spars have typically not used the buoyancy cans for tensioning. Instead, hydraulic tensioners have been used. Figure 13.9 shows a RAM type tensioner in which the hydraulic piston rods are in compression. This type of tensioner requires that the riser remain aligned with the hull, as it is on a spar. More traditional tensioner cassettes with the rods in tension have also been used (Genesis Spar).

The use of tensioners for drilling eliminates the need for the drilling riser slot to be increased in diameter to accommodate larger buoyancy cans that might be required for drilling.

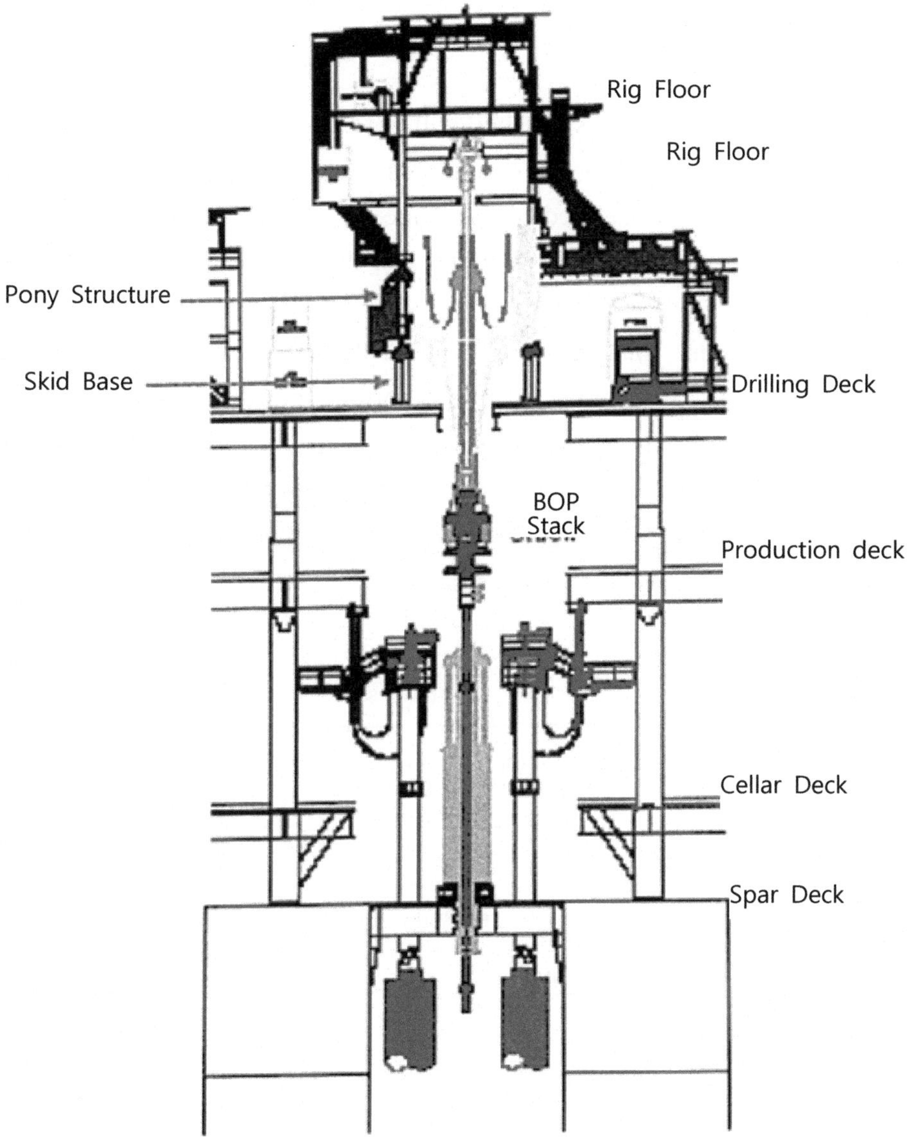

▲ **Figure 13.9** Tensioning of a spar drilling riser (Bates et al, 2000)

13.5 Spar Mooring

The mooring system for the spars built till date consists of a chain-wire-chain taut catenary system similar to the one shown in figure 13.10 for the Neptune Spar. A "taut" mooring is defined here as one in which the anchor loads have an uplift component for all load conditions, i.e. the anchor chain or wire never lies on the seabed.

The spar motions are small enough, even in the 100-year hurricane, that the taut system can be used without synthetic mooring lines. The taut system saves a considerable length of wire and chain needed for a conventional catenary mooring.

The platform chain is tensioned using chain jacks or windlasses, which are installed on the periphery of the upper deck of the hull (the "spar deck"). Figure 13.4 shows an example of the arrangement of this deck.

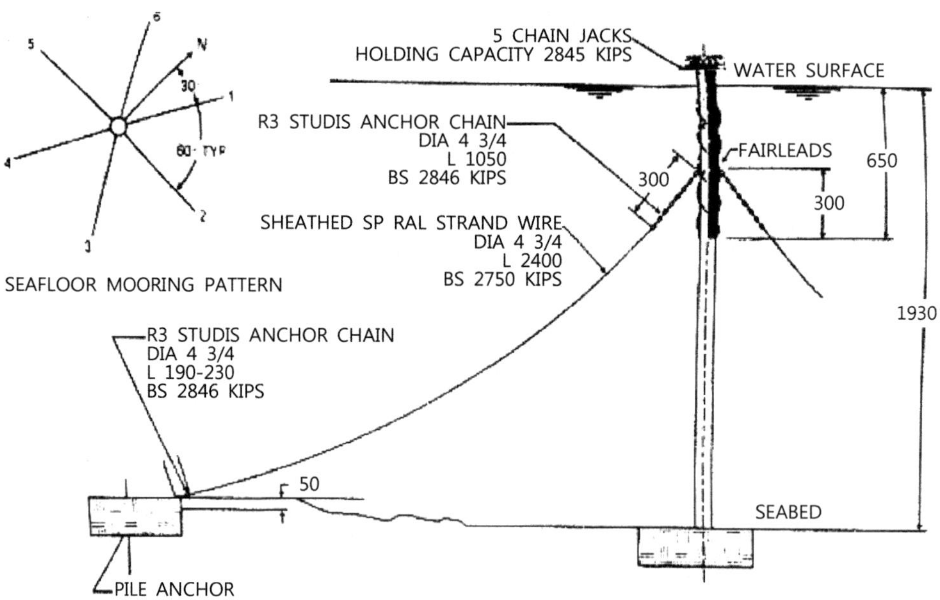

▲ **Figure 13.10** Neptune spar mooring system (Vardeman et al, 1997)

The chain runs from the chain jack to a fairlead, which is located from up to 350ft below the mean waterline. The length of platform chain is determined by the amount chain, which needs to be pulled in or paid out to maneuver the spar.

The midsection of the mooring system consists of a spiral strand wire rope or polyester line. For long life the steel strand is typically sheathed with a urethane coating. The lower end is attached to a length of anchor chain. The length of chain and the mooring tension is selected so that the wire will not make contact with the sea bottom except under the most extreme conditions.

The anchor chain is connected to a piled anchor which can sustain uplift and lateral loads. The pile padeye is usually about 50ft below the mudline, so that the bending moment from the mooring forces is minimized.

The total scope of the taut moor from the fairlead to the anchor is typically about 1.5 times the water depth.

Reference

Subrata Chakrabarti, Handbook of offshore Engineering, vol1, pp29-30, pp.536-545, Elsvier, 2003

Home Work Problem

1. Find the meaning of the following words or phrases by consulting the dictionary.

taut	offload	oceanography	drift
clump	hul	ambient	be made up of
consist of	fairlead	emanate	cofferdam
boom	upending	result in	stem
tensioner	cassette	synthetic	windlass
periphery	strand	lateral	padeye

2. Put T or F for the following sentences whether these are true or false, respectively, according to the facts in the passage of this unit.

- In the mid–eighties, Shell installed an oil storage and offloading Spar at Brent Field, in the North Sea.
- The Neptune Spar was installed in 1985 by Oryx energy Company (now Kerr McGee) and CNG.
- A third generation "cell" Spar was introduced in 2004.
- The "truss" Spars are the oldest type of Spars.
- Because of the length of a Spar, the Spar hull cannot be towed upright.
- Flip was put into service in 1965 and was used primarily for ocean acoustic measurements.

- The Spar, along with the Tension Leg Platform, is the only "floating" platform which up to 2003 has been used for wet trees.

3. Answer the following questions.

- Please name three types of production spars.
- What is a "taut" mooring? Please define it.
- What is the world's first production Spar?
- What is the function of the soft tank?
- What is the name of the first significant Spar?
- What is the name of the flare Spa constructed off West Africa in 1992?
- What is the name of the Japanese company which, in the early 1960s, installed a Spar off the coast of Japan to carry a microwave relay station?

GLOSSARY OF TERMINOLOGY AND EXPRESSIONS

While every effort has been made to fully explain the terminology and expressions used within the text of the book, there follows a list of terms designed to further clarify the more ambiguous references, and to provide a brief definition of some of the more frequently encountered expressions.

Accommodation space - Any room used for eating, sleeping, cooking, laundry or recreation, or as an office or sick bay, and any corridor giving access to any of these rooms, and any storeroom in the vicinity of any of these rooms.

Accumulator - A pressure vessel charged with nitrogen gas and used to store hydraulic fluid under pressure for the operation of hydraulic valve actuators.

Actuator - A mechanical device for the remote or automatic operation of a valve or choke

Annulus - Space between concentric casing strings.

Associated gas - Naturally occurring reservoir gas found in association with oil, either dissolved in the oil or found as a cap or pocket of free gas above the oil.

Azimuthing thruster - Rotatable ducted propeller used in conjunction with a dynamic positioning system to enable ships to maintain position without the use of anchors.

Block - The subdivided areas of the sea for the purpose of licensing to a company for exploration or production rights. In the UK a block is one thirtieth of a quadrant (one degree by one degree) and is approximately 200-250 square kilometers.

Blowout - Uncontrolled release of well fluids from the well bore during drilling operations.

Bell nipple - Receptacle attached to the top of the blowout preventer or marine drilling riser which directs the drilling mud returns to the shale shaker or mud pits.

BOP - Device attached to the casing head during drilling operations that allows the well to be sealed to confine the well fluids in the well bore.

Brown book - The Department of Trade and Industry annual publication contains a host of facts and figures relevant to oil and gas production in the UK and is available from Her Majesty's Stationary Office.

Bubble point - Point at which dissolved gasses begin to vaporize from a liquid. Dependent on temperature, pressure and gas/liquid composition.

Caisson - Length of pipe extending vertically downwards from an installation into the sea as a means of disposing of waste waters, or for the location of a sea water pump.

Casing - Pipe used to line and seal the well and prevent collapse off the borehole. A number of casing strings (lengths) are used in ever decreasing diameters.

Cathodic protection - Corrosion protection system which relies on sacrificial anodes or impressed current to protect submerged steel components from corrosion by electrolytic action.

Condensate - Volatile liquid consisting of the heavier hydrocarbon factions that condense out of the gas as it leaves the well, a mixture of pentanes and higher hydrocarbons.

Conductor - The first and largest diameter pipe to be inserted (spudded) into the seabed when drilling a well. It keeps the hole open, provides a return passage for the drilling mud and supports the subsequent casing strings.

Choke - A valve like device with a fixed or variable aperture specifically intended to regulate the flow of fluids.

Christmas Tree - An assembly of valves attached to the wellhead and used to control well production.

Crossover - Item used to connect one component to another differing in size, thread type or pressure rating.

Dew point - Temperature at which liquids condense from a gas.

Duster - Dry well drilled during exploration.

Dynamic positioning - Satellite monitoring system used to control the action of thruster propellers to maintain a vessel on location without deploying anchors.

Emergency shutdown valve (ESDV) - An automatically operated normally open valve used for isolating a subsea pipeline.

Enhanced oil recovery - Means used to assist in the extraction of oil either by installing equipment into the production tubing or by injecting water or gas into the reservoir.

Fire loop - A pneumatic control line containing temperature sensing elements (fusible plugs, synthetic tubing, etc.) which, when activated, will initiate a platform shutdown.

Fiscal metering - The measurement of oil, gas or condensate flow rate for taxation purposes.

Flare stack - Elevated tower containing a pipe used for the discharge and burning of waste gases.

Flash drum - Pressure vessel used to lower the pressure of oils and other liquids involved with the production process in order to encourage the vaporization of dissolved gases.

Flowline - Piping which directs well fluids from wellheads to manifold or first process vessel.

Fluid - A generic term meaning a gas, vapor, liquid or combinations thereof.

Grout - Mixture of cement and water (no sand) used to secure and seal attachments such as piles into jacket legs.

Header - That part of a manifold which directs fluid to a specific process system.

Hydrate - Solid, ice like compound consisting of molecules of water and hydrocarbon gasses.

Hydro-cyclone - Separation device utilizing centrifuging principles to remove oils from water, or as a multi-cyclone to remove liquids and solids from a gas stream.

Heat affected zone (HAZ) - That portion of the base metal which has not been melted, but whose mechanical properties or microstructure have been affected by the heat generated during the welding process.

Intrinsically safe - Electrical equipment which is incapable of igniting a flammable gas mixture or combustible materials.

Installation - May be fixed or mobile and used directly or indirectly for the exploration or production of mineral resources.

Installation, fixed - A fixed offshore structure involved in the production of oil or gas which may be constructed of steel or concrete. Term used frequently in the UK to describe an offshore installation.

Jacket - Steel support framework used to support platform topsides.

J-T valve - Throttle valve used to reduce the pressure and temperature of a gas stream, associated with the NGL removal process.

Knock out - Removal of liquids from a gas stream within a pressure vessel.

Lower explosive limit (LEL) - The lowest concentration by volume of combustible gases in mixture with air that can be ignited at ambient temperature conditions.

LNG - Liquefied natural gas, gaseous at ambient temperatures and pressures but held in the liquid state by very low temperatures to facilitate storage and transportation in insulated vessels.

LPG - Liquefied petroleum gas, essentially propane and butane held in the liquid state under pressure to facilitate storage and transportation.

Manifold - An assembly of pipes, valves and fittings by which fluid from one or more sources is selectively directed to various process systems.

Marine drilling riser - Pipe extending from the blowout preventer on the seabed to the drilling rig on the surface, to permit the return of the drilling mud.

Microwave - High frequency multi channel radio communications system designed to carry information between two points linked by line of sight transmission.

Mobile installation - One which can be moved from place to place without major dismantling or modification.

Module - Self contained liftable package forming part of the topside facilities of an offshore installation, e.g. accommodation module, compressor module, drilling module etc.

Multiphase - Practice of flowing unstabilized well fluids (oil with high gas content) in a single pipeline by boosting the pressure to prevent vaporization of the dissolved gasses.

Natural gas - Hydrocarbon gas occurring naturally from underground reservoirs both on and offshore.

NGL - Natural gas liquids, a mixture of hydrocarbon liquids which include butane and ethane obtained from natural gas. May be produced from condensate reservoirs but more probably produced as a byproduct of oil production.

Nipple - A section of threaded or socket welded pipe used as an appurtenance that is less than 12 inches in length. Often used to describe any short length of open ended pipe.

Nozzle - Flanged inlet or outlet connection on a pressure vessel.

Mudline - Seabed.

Packer - Device for sealing one casing string from another, or from the production tubing.

Pedestal - Large diameter, vertical tube or tub onto which a crane is attached.

Photogrammetric - The use of still photography to capture dimensional information for transposing into drawings.

Pig - Spherical device inserted into a gas subsea pipeline to sweep the line of deposits of rust, scale and condensed liquids. May also be used to clean oil pipelines of wax and may be "intelligent", that is, containing measuring and inspection equipment.

Pig trap - A pressure chamber permitting the entry or removal of equipment into the subsea pipeline, normally pigs.

Pipeline - Piping used to convey fluids between platforms or between a platform and a shore facility.

Pipe spool - Single length of pipe with flanged ends.

Platform, offshore - A fixed offshore structure involved in the production of oil or gas which may be constructed of steel or concrete. Term used frequently by Americans to describe an offshore installation.

Pressure vessel - Container, normally cylindrical used to contain internal or occasionally external pressure.

Produced water - Formation water removed from the oil and gas in the process pressure vessels.

Production separator - Main process vessel used primarily for the separation of gas, oil (and condensate) and water.

Production tubing - Pipe used in wells to conduct fluid from the producing formation into the Christmas tree. Unlike the casing the tubing is designed to be replaced during the life of the well, if required.

Purge - Maintain gas flow in an over rich, or lean concentration so as to avoid the buildup of oxygen and an explosive mixture.

Quality assurance - A sequence of planned and systematic actions necessary to provide adequate confidence that a product or service will satisfy given requirements of quality.

Quality control - The operational techniques and activities that are used to ensure that a quality product or service will be produced.

Rig - A term normally associated with drilling equipment, that is to say a drilling rig. Also a slang term used extensively to describe any of the structures and vessels associated with oil and gas exploration and production.

Riser - The vertical portion of a subsea pipeline (including the bottom bend) arriving on or departing from a platform.

Rock dumping - Deposition of rocks onto subsea pipelines to provide protection against anchors and fisherman's nets when burying the pipe is impractical. Rocks and gravel may also be dumped around subsea wellheads and jacket legs to repair scour damage.

Seismic survey - The use of artificially generated sound waves to determine the type of rock formations below the ground or under the sea bed by monitoring the reflected sound wave signals.

Scour - Removal of the seabed in the vicinity of a jacket, subsea wellhead or pipeline by tidal action.

Scrubber - Pressure vessel containing equipment designed to remove or scrub liquids from a gas stream.

Shuttle tanker - Moderate sized oil tanker used to transport oil from larger vessels into port.

Skid - Steel framework used to contain equipment, may be transportable.

Stress corrosion cracking - The cracking which results from a combination of stress and corrosion.

Slew ring bearing - Large ball or roller bearing which connects crane to pedestal and permits rotation.

Slug - An accumulation of water (may also be sand or condensate) in a gas pipeline.

Spudding - A term used to describe the insertion of the conductor into the seabed when drilling a well. May also be used to describe the process of setting the legs of a jack-up into the seabed.

Stinger - Tubular steel support frame attached to the stern of a pipelay barge to control the bending of the pipe as it enters the water.

Stripping - The removal or replacement of drill pipe or tubing strings from a well under pressure using a stripping BOP.

Stripping gas - Gas, normally process gas used to assist in the purification of a liquid by reducing the partial pressure of gaseous contaminants to encourage vaporization.

Swabbing - The lowering of the hydrostatic pressure in the hole due to the upward movement of the drill pipe and/or tools. Also the use of wireline equipment to clean a well by scooping out liquids.

Sulphide stress cracking - Cracking of metallic materials due to exposure to fluids containing hydrogen sulphide.

Telemetry - System for the collection, collation and transmission of information to a remote source using radio, satellite, fiber optics or cable links. Also associated with the remote control of process equipment.

Third party gas - Term used to describe gas sold direct from oil company to parties other than British Gas, the previous monopoly holders in the UK.

Throttle - Regulation of fluid flow by a throttling valve or fixed orifice.

Topsides - Upper part of a fixed installation which sits on top of the jacket and consists of the decks, accommodation and process equipment.

Vam - Trade name for casing thread produced by the Vallourec company of France.

Vent - A pipe or fitting on a vessel that can be opened to the atmosphere.

Vent stack - Open ended pipe and support framework used to discharge vapors into the atmosphere at a safe location above the installation without combustion.

Wellhead - Permanent equipment used to secure and seal the casings and production tubing and to provide a mounting place for the Christmas trees.

Wireline - Equipment used to introduce tools into the well bore under pressure.

Workover - Re-entry into a completed well for modification or repair work.

Workover rig - Normally a smaller, portable version of the main drilling derrick which can be used to carry out work over operations on installations which do not have a permanent derrick.

Reference

Angus Mather, Offshore Engineering–An Introduction, pp.xix–xxiv, Witherby & Co., 1995

PRACTICAL ENGLISH IN OFFSHORE PLANT ENGINEERING

2014년 8월 20일 초판 인쇄
2015년 11월 05일 2 쇄 발행

저 자 김동혁
발 행 인 송기수
발 행 처 도서출판 GS인터비전
편 집 처 도서출판 GS인터비전
인 쇄 처 대일문예사
등록번호 제 313-2011-120 호
I S B N 979-11-5576-046-8(93740)

주 소 서울 은평구 증산로15길 69, 2층
전 화 02-976-7898, 02-3272-7898.
팩 스 02-6468-7898
홈페이지 gsintervision.co.kr
E-Mail gsinter7@gmail.com

정 가 17,000원